# Victory, Defeat, or Draw

"David Rodman's *Victory, Defeat, or Draw* presents a detailed account of eight critical battles waged by the Israel Defense Forces during its long and storied history. Rodman delves deeply into these clashes, and his conclusions are provocative and might make you rethink why armed forces win, lose, or tie in modern warfare."

Kenneth M. Pollack, author of *Armies of Sand: The Past, Present, and Future of Arab Military Effectiveness*

"David Rodman's excellent book is timely, as the question of victory or defeat in battle remains a pressing one for both generals and politicians in the 21st century. Using cases from the Israel Defense Forces, Rodman's research demonstrates that the importance of such classic variables as planning, intelligence collection, and combined arms transcends time and place. This book is essential reading for anyone who wants to learn about how battles are fought and won."

Eitan Shamir, senior research associate, BESA Center for Strategic Studies, Bar-Ilan University, and author of *Transforming Command*

"David Rodman's *Victory, Defeat, or Draw* offers a timely and sophisticated review of the complex conditions that determined the outcomes of battles fought by the Israel Defense Forces from the War of Independence to the First Lebanon War."

Alon Kadish, professor emeritus of history, The Hebrew University, and author of *The British Army in Palestine and the 1948 War*

# Victory, Defeat, or Draw

Battlefield Decision in the Arab–Israeli Conflict, 1948–1982

DAVID RODMAN

sussex
ACADEMIC
PRESS
*Brighton • Chicago • Toronto*

2 4 6 8 10 9 7 5 3

*First published in hardcover 2019, reprinted in paperback 2020, in Great Britain by*
SUSSEX ACADEMIC PRESS
PO Box 139
Eastbourne BN24 9BP

SUSSEX ACADEMIC PRESS
Independent Publishers Group
814 N. Franklin Street, Chicago, IL 60610



*British Library Cataloguing in Publication Data*
A CIP catalogue record for this book is available from the British Library.

*Library of Congress Cataloging-in-Publication Data*
Names: Rodman, David, author.
Title: Combined arms warfare in Israeli military history : from the War of Independence to Operation Protective Edge / David Rodman.
Description: Brighton ; Chicago ; Toronto : Sussex Academic Press, [2019] | Includes bibliographical references and index.
Identifiers: LCCN 2018046022 | ISBN 9781845199678 (hbk : alk. paper) | ISBN 9781845199807 (pbk : alk. paper)
Subjects: LCSH: Israel—History, Military | Israel. Tseva haganah le-Yisrael—History | Unified operations (Military science).
Classification: LCC DS119.2 .R63 2019 | DDC 355.4/6095694—dc23
LC record available at https://lccn.loc.gov/2018046022

MIX
Paper | Supporting responsible forestry
FSC® C013604

Typeset & designed by Sussex Academic Press, Brighton & Eastbourne.
Printed and bound by CPI Group (UK) Ltd, Croydon, CR0 4YY

# Contents

# List of Abbreviations

| | |
|---|---|
| AAA | Antiaircraft Artillery. |
| AEW | Airborne Electronic Warfare. |
| AFV | Armored Fighting Vehicle. |
| APC | Armored Personnel Carrier. |
| ARM | Antiradiation Missile. |
| AWACS | Airborne Warning and Control System. |
| $C^2$ | Command and Control. |
| $C^3I$ | Command, Control, Communications, and Intelligence. |
| CAP | Combat Air Patrol. |
| CAS | Close Air Support. |
| CAW | Combined Arms Warfare. |
| CBU | Cluster Bomb Unit. |
| CIA | Central Intelligence Agency. |
| CO | Commanding Officer. |
| CP | Command Post. |
| EW | Electronic Warfare. |
| FPB | Fast Patrol Boat. |
| GCI | Ground Control Intercept. |
| GHQ | General Headquarters. |
| GP | General Purpose. |
| HQ | Headquarters. |
| HUMINT | Human Intelligence. |
| IADS | Integrated Air Defense System. |
| IAF | Israel Air Force. |
| IDF | Israel Defense Forces. |
| IED | Improvised Explosive Device. |
| IN | Israel Navy. |
| KIA | Killed in Action. |
| LIC | Low-intensity Conflict. |
| MOSSAD | Israeli Foreign Intelligence Service. |
| NATO | North Atlantic Treaty Organization. |
| NCO | Noncommissioned Officer. |
| ORBAT | Order of Battle. |

| | |
|---|---|
| PFLP | Popular Front for the Liberation of Palestine. |
| PGM | Precision-guided Munition. |
| PHOTINT | Photographic Intelligence. |
| PLO | Palestine Liberation Organization. |
| RMA | Revolution in Military Affairs. |
| SAM | Surface-to-air missile. |
| SEAD | Suppression of Enemy Air Defense. |
| SIGINT | Signals Intelligence. |
| SOF | Special Operations Force(s). |
| SPEC OP | Special Operation. |
| SSM | Surface-to-surface Missile. |
| UAV | Unmanned Aerial Vehicle. |
| UN | United Nations. |
| WMD | Weapons of Mass Destruction. |
| WP | Warsaw Pact. |

Note: The abbreviations AL, EG, and IS in connection with numbered military formations indicate the Jordanian, Egyptian, and Israeli armies, respectively.

# List of Illustrations

**Photo section (pages 55–58)**

# Preface

Three outcomes are possible on the battlefield: victory, defeat, or draw. An adversary may defeat or be defeated by its adversary, or neither of the two may emerge victorious or vanquished. Observers of military history have long tried to identify the variables that determine victory, defeat, or draw. While most would certainly acknowledge that decisions on the battlefield are dictated by a combination of variables rather than by a lone circumstance, many observers nevertheless tend to stress a single variable—for example, the number of fighting men and fighting machines deployed by the adversaries, or the operational doctrines employed by the opposing forces—as far more significant to the explanations of these decisions than other variables.

This book, in contrast, takes a multicausal approach to the question of victory, defeat, or draw, proposing that a combination of six organizational, matériel, and environmental variables are pivotal to the explanation of decision on the battlefield. Using the extensive history of the Israel Defense Forces (IDF), the book examines a sample of eight battles across the Arab–Israeli conflict from 1948 to 1982 in order to determine the collective impact of the six variables on the outcomes of these battles, concluding that this basket of variables captures much of the explanation behind victory, defeat, or draw on the battlefield, at least insofar as concerns the record of the IDF.

Before moving on to the research itself, it is first necessary to acknowledge SAGE Publishing and the Israel Government Press Office/Israel Defense Forces for granting permission to use their copyrighted text and photographs, respectively, in this monograph. Chapter 5 is a revised and adapted version of David Rodman, "Turning Point in the Sinai: The Great Armoured Battle of 14 October 1973," *War in History* (forthcoming). The online version of this article may be found at <journals.sagepub.com/home/wih>. The author would also like to thank the two anonymous reviewers at *War in History* for their very encouraging and helpful comments about the first draft of this article.

David Rodman
Dix Hills, New York

# INTRODUCTION
# Victory, Defeat, or Draw

That warfare is among the most complex of human endeavors is not in dispute. That the interactions of many variables determine which adversary wins and which one loses a battle or a war, or whether neither adversary wins nor loses, is not in dispute, either. It is relatively easy to observe in retrospect which adversary won and which one lost a battle or a war, or whether neither adversary won nor lost; however, it is certainly not nearly as easy to determine why the victor won and why the vanquished lost, or why neither adversary won nor lost. Indeed, the very complexity of warfare has led to the rise of various schools of thought with respect to the question of which class of variable holds the greatest importance in determining the outcome of a battle or a war.

One school of thought emphasizes the importance of culture—that is, a collection of national attributes, including the type of government (democratic versus non-democratic) and the level of civic-mindedness of the population. For adherents of this school, such variables explain why polities with Western cultures have tended to prevail over polities with non-Western cultures.[1] A second school of thought emphasizes the importance of the manpower component of military organizations—that is, such variables as the level of training, motivation, and experience of rank-and-file soldiers; the training, motivation, and experience of the officer corps; and the strategic-, operational-, and tactical-level doctrines that govern actions on the battlefield. For adherents of this school, "human factors" constitute the overriding determinant of victory or defeat on the battlefield.[2] A third school of thought, predictably, emphasizes the importance of the matériel component of military organizations—that is, such variables as the numbers of fighting men and fighting machines, as well as the technological sophistication of the fighting men and fighting machines, at the disposal of these organizations. For adherents of this school, "non-human factors" constitute the difference makers on the battlefield.[3] Yet a fourth school of thought emphasizes the importance of the environment—that is, such variables as terrain, climate, and weather. For adherents of this school, victory typically belongs to the adversary better prepared to cope with Mother Nature.[4] Victories that cannot be ascribed to one of these four classes of variable are often chalked up to dumb luck.

Rather than emphasize a single class of variable, a more fruitful approach to the question of what leads to victory, defeat, or draw on the battlefield is to call upon elements of different classes. Specifically, the historical record of warfare across both time and space indicates clearly that a limited set of variables exercises a particularly prominent role in explanations of victory, defeat, or draw on the battlefield. These variables include: the effectiveness of operational planning prior to the battle, the effectiveness of intelligence collection and reconnaissance activities prior to and during the battle; the effectiveness of the exploitation of the terrain of the battlefield; the effectiveness of the forces (with respect to their quantity and quality) deployed on the battlefield; the effectiveness of command and control ($C^2$) during the battle; and the effectiveness of the practice of combined arms warfare (CAW) during the battle.

The present monograph applies this more nuanced approach to the case of the Israel Defense Forces (IDF).

## Why the Case of the Israel Defense Forces

The State of Israel emerged onto the world stage in the middle of a war, and it has been entangled in one type or another of armed conflict almost continuously throughout its existence. From its formal establishment in 1948 to the early 1980s, the country fought six interstate wars against Arab states. These wars included: the 1947–49 War of Independence; the 1956 Sinai Campaign; the 1967 Six-Day War; the 1969–70 War of Attrition; the 1973 Yom Kippur War; and the 1982 Lebanon War. During the first two decades of the twenty-first century, it fought three asymmetrical wars against Arab insurgent/terrorist groups. These wars included: the 2006 Second Lebanon War; the 2008–9 Gaza War (Operation Cast Lead); and the 2014 Gaza War (Operation Protective Edge). During its first 70 years, then, Israel averaged more than one war per decade. And, between these wars, the country has been involved in various types of lower-level hostilities—for example, border skirmishing—with its numerous adversaries.

Regardless of whether it has been engaged in interstate war, asymmetrical war, or lower-level hostilities, Israel has amassed a rather remarkable record of success in warfare. Indeed, it achieved clear-cut battlefield victories in five of the six interstate wars that it has fought to date, while the sixth war ended in a stalemate. It also achieved clear-cut battlefield victories in two of the asymmetrical wars that it has fought to date, while the third war ended in a more limited triumph. The country's overall rate of success on the battlefield in lower-level hostilities has been equally impressive. Though it has certainly suffered military setbacks from time to time, Israel's record of success in warfare is probably unequaled by that of any other country in the post-Second World War period.

The country's extensive military history represents fertile ground for conducting an assessment of the extent to which the aforementioned variables do indeed dictate victory, defeat, or draw. To this end, a focused comparison of eight pivotal battles—six fought during interstate wars and two fought during lower-level hostilities—is sufficient to reach some tentative conclusions in respect of cause and effect on the battlefield. These battles, in order of appearance in the text, are: the Battle of Latrun during the War of Independence; the First Battle of Abu Ageila/Um Katef during the Sinai Campaign; the Second Battle of Abu Ageila/Um Katef during the Six-Day War; Operation Blossom during the War of Attrition; the Battle of October 14 (Sinai front) during the Yom Kippur War; Operation Mole Cricket 19 during the Lebanon War; and Operations Hell and Spring of Youth during the period between the Six-Day War and the Yom Kippur War.

Six of these battles were primarily ground battles, while two were primarily air battles.[5] Seven of these engagements were offensive (i.e., initiated by the IDF), while the eighth was defensive (i.e., initiated by its adversary). Four of these battles were waged against the Egyptian army, one against the Jordanian Arab Legion, one against the Syrian army, one against Palestine Liberation Organization (PLO) forces, and one against a combined Arab Legion–PLO contingent. The mixed outcomes of battle (five victories for the IDF, two defeats, and one draw), the mixed types of battle (six ground and two air, seven offensive and one defensive, six in the context of interstate war and two outside this context), and the mixed adversaries in battle (Egyptians, Jordanians, Syrians, and Palestinians), along with the fact that one battle from each of Israel's six interstate wars and two battles outside of an interstate war context are under consideration, ensures that a reasonably random sample of pivotal battles fought by the IDF is represented here, which is a necessary basis for arriving at valid, albeit tentative, conclusions with respect to cause and effect on the battlefield.

## Definitions of the Variables

Before dipping into the battles themselves, however, it is first obligatory to define briefly the independent and dependent variables in general terms so that the concepts that lay behind them are not ambiguous. For the purpose of this research, all of the independent (or causal) variables are conceived of as trinary in character—that is, they have three possible "values." For these six variables, then, the values are " more effective," "equally effective," and "less effective." The dependent (or outcome) variable is also conceived of as trinary in character. For this variable, the values are "victory," "defeat," and "draw."

### *Dependent Variable*

The dependent variable—*the results obtained by the IDF*—is not as straightforward as might appear at first consideration. For the result of a battle to be coded as a victory for the IDF, the outcome of the battle had to promote the realization of the IDF's broader military objective(s) in the hostilities. For the result of a battle to be coded as a defeat for the IDF, the outcome of the battle had to impede the realization of the IDF's broader military objective(s) in the hostilities. And for the outcome of a battle to be coded as a draw for the IDF, the outcome of the battle had neither to promote nor to impede the realization of the IDF's broader military objective(s) in the hostilities. Victory, defeat, or draw, in short, is not conceived here solely as a function of what happened on the battlefield itself.

### *Independent Variables*

*Operational planning* is defined here as the advance preparations made for battle by each adversary—that is, each adversary's effort to assemble and deploy the appropriate forces in a manner designed to ensure that they are at an advantage vis-à-vis the opponent's forces and that their anticipated actions will contribute positively to the broader war effort. Methodical planning is generally more crucial in offensive than in defensive battles, as the former are generally more difficult to execute than the latter.

*Intelligence and reconnaissance* activities prior to and during battle are defined here as the timely collection, interpretation, and dissemination of information instrumental to prevailing on the battlefield. The more complete, accurate, and current the information available to an adversary, the more likely it is to be triumphant on the battlefield.

*Exploitation of terrain* is defined here as the extent to which each adversary takes advantage of the topographical features of the battlefield in order to improve its prospect of victory. The better use an adversary makes of the terrain, the better its prospect of victory on the battlefield.

*Quantity and quality of forces* is defined here as the numbers and capabilities of the fighting men and fighting machines deployed during battle. To win on the battlefield, an adversary typically requires not only that its forces possess superior capabilities to the ones possessed by its opponent, but also that it possesses sufficient numbers to overcome any quantitative advantage held by its opponent.

*$C^2$* is defined here as the management of forces during battle—that is, the coordinated employment of those forces in order to accomplish the desired battlefield objective(s). An adversary with a proficient $C^2$ system is more likely to emerge triumphant on the battlefield than one with a less competent $C^2$ system.

*CAW* is defined here as the integration of different arms—for example, armor, infantry, artillery, and engineers—on the battlefield in order to accomplish the desired military objective(s). The thinking behind CAW, which has been repeatedly verified through actual experience, is that the simultaneous employment of different arms acts as a "force multiplier" by allowing each arm to take advantage of its inherent strengths, while other arms compensate for its inherent weaknesses. An adversary that is proficient in the practice of CAW is more likely to win on the battlefield than one that is less proficient in the practice of CAW.

For the IDF to receive a grade of more effective in regard to a particular independent variable, its performance during a battle must have been better than the performance of its adversary in regard to that variable. Conversely, for the IDF to receive a grade of less effective in respect of an independent variable, its performance during a battle must have been worse than the performance of its adversary in regard to that variable. And, for the IDF to receive a grade of equally effective in regard to a particular variable, its performance during a battle must have been more or less equivalent to the performance of its adversary in regard to that variable.

Finally, the boundaries that separate the independent variables, it must be conceded, are rather permeable in certain instances—for example, $C^2$ is an integral part of CAW. Nevertheless, despite the evident overlap, it is still worthwhile to consider all of the variables individually in order to achieve as much descriptive and analytical clarity as possible, regardless of whether doing so is entirely sound from a methodological point of view.

## The Structure of the Research

The substantive chapters are structured in the same way. Each battle is first set in its larger historical context. This background is followed by a review of the prelude to the battle, a description of the battle itself, and then an analysis of the battle in light of the six independent variables. Finally, the broader ramifications of each battle are set forth. Only the final substantive chapter deviates from this structure in that it encompasses both battles that took place outside of the context of interstate war.

It should be stressed here that the descriptions of the battles profiled in this book are not meant to be comprehensive in scope. It is not the intention of the research presented here to recount in-depth the part played by each battalion, brigade, division, or squadron in this or that battle. Rather, the intention is to provide only enough description as to make possible an analysis of the battle in light of the independent variables. Moreover, while the general narratives of the battles are not in serious dispute, some of the particulars—for example, the part played by this or that unit in the hostilities, the numbers of fighting men and fighting machines committed to the hostilities, the number of fighting men and fighting machines lost in the hostilities, and the like—often vary from source to source, sometimes significantly. This monograph adopts those facts and figures that seem to be the most plausible available.

## One More Caveat

It should also be stressed that the research presented here is strictly preliminary in nature. This monograph certainly does not pretend to offer a rigorous "theory of victory" in respect of the IDF, let alone in regard to other national armies. To develop and test such a theory would require a formal model with very precisely defined variables (and, perhaps, additional relevant independent variables), as well as a set of logically consistent propositions that specify how all of these variables interact with one another. It would also require an impeccably random sample of battles, with some of them subjected to in-depth, focused comparisons and the rest subjected to systematic mathematical/statistical analysis. These conditions, obviously, do not apply to the research presented here. Rather, this research is meant simply to stimulate thought on the causes of victory, defeat, or draw, at least in regard to the IDF.

# 1

# The Battle of Latrun

Beginning in the late nineteenth century, Jews began to return to their ancestral homeland of Eretz Israel (Palestine) in growing numbers from the lands of their exile. The steady expansion of the Yishuv (Settlement), as the Jewish community referred to itself before the establishment of the State of Israel, generated increasing friction with the surrounding Arab population, which resented its presence. During the first decades of the Jewish return, marauding Arab bands intent on pillage and destruction constituted the primary threat to the Yishuv. After the British government issued the Balfour Declaration in 1917, a document that recognized the Jewish people's right to rebuild a "national home" in its ancestral homeland, however, Arab violence against the Yishuv became not only much more overtly political in character, but also much more organized, widespread, and murderous in nature.

By the early twentieth century, the Yishuv realized that it had no alternative but to set up a self-defense force in order to protect its members, as the Ottoman authorities who ruled Palestine at the time, like the local Arabs, were none-too-friendly to the Jewish community; consequently, during the first decade of the century, HaShomer (The Watchman) emerged to occupy this role, its most urgent task to guard Jewish agricultural villages from Arab marauders.[1] In the post-Balfour Declaration climate of Palestine, though, the Yishuv concluded that a small self-defense force such as HaShomer would not be sufficient to protect the community, even if the British authorities who had replaced the Turkish authorities of the defunct Ottoman Empire as Palestine's rulers after the First World War were more committed to countering anti-Jewish violence than their predecessors. Consequently, the Yishuv replaced HaShomer with Haganah (Defense), an organization that would evolve over the following decades into a rather large and efficient underground militia, which would also stand up an elite force known as the Palmach (an abbreviation for "strike companies") during the

Second World War.[2] The Haganah's policy of self-restraint in the face of Arab violence eventually spawned two dissident militias—the Irgun Zvai Le'umi (National Military Organization) and the smaller, more extreme Lohamei Herut Yisrael (Fighters for the Freedom of Israel), the former commonly referred to as the Irgun and the latter as Lehi—that adopted aggressive, tit-for-tat approaches in their responses to Arab attacks.[3]

While the British authorities engaged out of necessity in some low-key cooperation with the Haganah during the 1936–39 Arab Rebellion, as well as during the Second World War, they nevertheless considered it to be an unsanctioned militia.[4] Their attitude toward the organization hardened further once large-scale armed Jewish resistance to British rule commenced in the aftermath of the Second World War.[5] The British authorities adopted an even tougher line against the Irgun and Lehi, branding them as outright "terrorist" organizations. The Haganah, Irgun, and Lehi, therefore, always encountered great difficulty in acquiring arms, as the British authorities actively sought to impede the flow of weapons to these organizations. Indeed, at the start of the 1947–49 War of Independence, their arsenals contained only light mortars, machine guns, rifles, pistols, and improvised explosive devices (IEDs)—and all of these types of arms in very limited quantities. In terms of fighting machines, then, the Yishuv's military forces were woefully ill-equipped for full-scale war on the eve of the War of Independence.

The same holds true with respect to manpower. While the Haganah, Irgun, and Lehi had acquired a certain amount of organizational military experience during the course of clashes with Arab insurgents and British soldiers, this exposure to armed conflict occurred in the context of operations associated with insurgency/counterinsurgency, not full-scale war. Moreover, though some individual members of the Haganah had acquired their military experience in the context of full-scale war when they fought with the British army during the Second World War, some as part of a Jewish Brigade that saw action in Italy, they were too few in number to exercise a decisive influence on how the Haganah prepared its forces for battle.[6] On the eve of the War of Independence, the manpower of the Yishuv's military forces simply was not adequately prepared for the test of full-scale war.

The shortcomings of these forces in respect of full-scale war would become swiftly and readily apparent when the nascent Israel Defense Forces (IDF) attempted to capture Latrun from the Jordanian Arab Legion on several occasions.

## Prelude to the Battle of Latrun

Unable to govern effectively the territory of western Palestine after the Second World War—Great Britain had unilaterally detached the 80 percent of the mandate's territory that lay east of the Jordan River during the 1920s, eventually creating the Hashemite Kingdom of Transjordan (later shortened to the Hashemite Kingdom of Jordan), an exclusively Arab state—the British government turned the decision on the fate of the remainder of Palestine over to the United Nations (UN). In November 1947, this body passed a partition resolution dividing western Palestine into separate Jewish and Arab states upon the termination of the British mandate in May 1948. While the Yishuv accepted the UN's "two-state solution," the Arab world furiously rejected the resolution, vowing to abort the establishment of a Jewish state. And, while the Arab world immediately began to engage in international diplomacy to overturn the UN's decision, the local Palestinian Arab community—with the assistance of foreign Arab "volunteers"—simultaneously went to war against the Yishuv.

The War of Independence passed through two distinct phases: an intercommunal phase from late November 1947 through mid-May 1948, during which Jewish and Arab militias fought each other, and an interstate phase from mid-May 1948 until early 1949, during which the newly formed IDF fought the invading armies of surrounding Arab states.[7] The War of Independence, then, was partly a low-intensity conflict (LIC) and partly an interstate war.

At the outset of the intercommunal phase of the war, Jewish forces were on the defensive, Arab forces on the offensive. On the one hand, Jewish forces proved to be quite capable of defending Jewish towns, neighborhoods, and agricultural villages effectively against direct attacks by Arab forces. Not only did these attacks make scant headway, but also they were frequently repulsed with heavy losses inflicted on the Arab attackers. On the other hand, Jewish forces were not nearly as effective in keeping open the vital road network that connected Jewish towns, neighborhoods, and agricultural villages. Arab forces often proved to be able to cut vital supply lines that linked Jewish population centers to one another, putting at risk not only the territorial integrity of the Yishuv itself, but also the very survival of outlying towns, neighborhoods, and agricultural villages whose continued existence depended on a secure road network.

Consequently, the Yishuv's leadership determined that Jewish forces must go over to the offensive in order to keep open the road network. Jewish forces put this decision into practice in the form of

Operation Nachshon and Plan D.[8] These operations had two basic objectives: to relieve pressure immediately on isolated portions of the Yishuv by taking control of the road network, and to capture as much territory assigned to the Yishuv under the terms of the UN partition resolution as feasible before the termination of the British mandate—in other words, to put the Yishuv in the strongest possible position to resist the forthcoming Arab invasion of the fledgling State of Israel. Throughout April and early May 1948, Jewish forces made significant headway in accomplishing these objectives. Notwithstanding the fact that Jewish forces had clearly gained the upper hand against local Arab forces before the termination of the British mandate, the War of Independence was about to enter its most difficult phase for the Yishuv.

On May 14, the last British forces withdrew from western Palestine, formally ending 30 years of British rule there. Israel proclaimed its independence on the same day. Immediately upon its declaration of independence, the armies of five surrounding Arab states—Egypt, Jordan, Iraq, Syria, and Lebanon—simultaneously invaded the new country, assaulting its territory from the south, east, and north.

Upon independence, the Haganah became the IDF. Irgun and Lehi forces would eventually be folded into the IDF as well. At the outset of the interstate phase of the War of Independence, the IDF order of battle (ORBAT) consisted of 40,000–50,000 soldiers (including women), of whom approximately half, organized into nine brigades, were suitable for field operations, while the other half were suitable simply for static defense.[9] The IDF possessed enough light weapons—mortars, machine guns, rifles, pistols, and IEDs—to equip only about half of its manpower. Its arsenal did not contain any heavy weapons of significance: no fighter-bombers, no tanks, no purpose-built armored vehicles, no artillery, and no warships. Such arms would only begin to trickle into its arsenal shortly after independence. The Yishuv had acquired these weapons abroad earlier from countries in both the Eastern bloc (through official channels) and the Western bloc (through unofficial channels), but these weapons could not be brought into western Palestine while the British remained in control.

For its part, the coalition of Arab armies and local militias arrayed against the IDF had similar numbers of soldiers suitable for both field operations and static defense.[10] The combined Arab forces, contrary to popular thought, never had a major advantage over the IDF in respect of the number of soldiers at their disposal. Rather, the coalition's major advantage over the IDF at the outset of the interstate phase of the War of Independence lay in its arsenal, which boasted substantial quantities of heavy weapons, including fighter-bombers, tanks, armored vehicles, and artillery, some of these arms being among the most modern available at the time.

## The Battle of Latrun

During the early part of the interstate phase of the War of Independence, the Arab coalition succeeded in capturing stretches of territory in the north, center, and south of Israel, primarily as a result of the coalition's possession of superior firepower. Nevertheless, on the whole, the IDF fought rather effectively on defense, eventually containing the Arab coalition forces before they could overrun vital Jewish areas. When the IDF went on offense during the early part of the interstate phase of the war, however, it often did not fare nearly as well as it did on defense. The Battle of Latrun is a very prominent case in point. Indeed, nowhere did the IDF perform more poorly on offense throughout the entire War of Independence than at the disastrous Battle of Latrun.[11]

The strategic significance of Latrun for Israel lay in its command of the road from Tel Aviv to Jerusalem. During their rule in Palestine, the British had built an imposing police fort atop a hill in the area in order to ensure that they alone could control the traffic on this road.[12] At the end of April, the British had withdrawn their forces from this fort. Local Arab militiamen promptly seized the empty position, eventually turning it over to the Hashemite Kingdom's Arab Legion upon the final British withdrawal from Palestine in mid-May. Jewish Jerusalem (i.e., western Jerusalem), already close to collapse as the result of a long-standing Arab siege, was now entirely cut off from the rest of Israel. To relieve the siege on Jerusalem, therefore, the Israeli government ordered the IDF to capture Latrun and reopen the road to the city.

To prevent the IDF from doing so, the Jordanians moved the 2nd (AL) and 4th (AL) Infantry Battalions into positions in and around Latrun.[13] The 4th (AL) Infantry Battalion anchored itself on the police fort and its neighboring village, while the 2nd (AL) Infantry Battalion anchored itself in some nearby villages. These infantry battalions had at their disposal modern armored fighting vehicles (AFVs) and artillery, and they could call upon the support of a large number of local militiamen.

For their part, the Israelis initially mustered the 72nd (IS) and 73rd (IS) Infantry Battalions of the newly formed 7th (IS) Infantry Brigade (later renamed and reequipped as the 7th (IS) Armored Brigade), as well as the 32nd (IS) Infantry Battalion of the 3rd (IS) Infantry Brigade, for the attack on Latrun.[14] The third battalion of the 7th (IS) Infantry Brigade had begun to absorb the first armored personnel carriers (APCs) brought in from abroad, but this battalion was not initially present at Latrun. The only "heavy weapons" that the IDF could muster at first to support these infantry forces consisted of a couple of ancient, pre-First World War French mountain guns.

Called Operation Bin-Nun, the first Israeli attempt to capture Latrun—the IDF would make at least four attempts over the following weeks—began during the early morning hours of May 25.[15] It went horribly wrong from the start. What the IDF had intended to be a night attack under the cover of both darkness and an artillery barrage ended up being a daylight attack without any such cover. The "artillery barrage" consisted of a small number of shells hurled haphazardly toward Arab Legion positions by a mere two guns, which served only to give Jordanian forces an early warning that IDF infantry forces were heading toward them. Moreover, the Israeli infantry forces slated for the attack arrived at their jump off points hours behind schedule, so the sun was already up by the time that they approached Arab Legion positions.

Under these circumstances, the 32nd (IS) Infantry Battalion, clearly visible to Arab Legion forces from their commanding positions on the high ground as it advanced toward them in daylight, was simply shot to pieces by accurate artillery, machine gun, and small arms fire before it could even get close to its objective, and it soon retreated in utter disarray. The 72nd (IS) Infantry Battalion—the 73rd (IS) Infantry Battalion was held in reserve—suffered much the same fate. Caught in the flank by accurate artillery, machine gun, and small arms fire from Arab Legion positions of which it was not aware, it, too, incurred heavy casualties, and it, too, quickly retreated in utter disarray.

The first attempt to take Latrun had essentially ended before it had really got underway. More than 70 IDF soldiers perished in the field, and IDF forces left behind a considerable quantity of small arms.[16] Arab Legion forces, including local militiamen, in contrast, lost only a handful of men in the fighting.

Despite the terrible beating it had taken during the first attempt to capture Latrun, the IDF launched a second assault five days later, on May 30.[17] Known as Operation Bin-Nun B, this assault, like the first one, unfolded as a two-pronged attack led by the 7th (IS) Infantry Brigade. This time around, elements of the 72nd (IS) Infantry Battalion of the 7th (IS) Infantry Brigade, along with elements of the 52nd (IS) Infantry Battalion of the 5th (IS) Infantry Brigade, which had since replaced the badly battered 32nd (IS) Infantry Battalion of the 3rd (IS) Infantry Brigade, were to attack a number of Arab villages in the Latrun area, while elements of the 71st and 73rd (IS) Infantry Battalions were to attack the Latrun police fort and neighboring village.

The second assault on Latrun fared little better than the first one. Arab Legion forces spotted the advance of the 72nd (IS) Infantry Battalion, hitting it with accurate artillery, machine gun, and small arms fire that inflicted considerable casualties, causing the battalion to retreat in disarray. As a consequence of the 72nd (IS) Infantry Battalion's plight, the 52nd (IS) Infantry Battalion retreated as well, having taken only a minor part in the fighting.[18]

Simultaneously, under the cover of an accurate artillery barrage, the 71st and 73rd (IS) Infantry Battalions, which this time around also had some APCs and other AFVs attached to their ranks, made some progress in their attack, even penetrating the defenses of the police fort with flamethrowers and explosives. While IDF mechanized and engineering forces had done their jobs in breaching Arab Legion defenses, however, the IDF nonmechanized infantry forces failed to exploit the breach with a follow-up attack, allowing the Jordanian defenders to regroup and repel all of the Israeli forces.[19]

Though IDF forces had acquitted themselves somewhat better than in the first attempt to capture Latrun, the second attempt had nevertheless cost the IDF another 40 plus soldiers killed in battle.[20] Arab Legion losses, on the other hand, were once again a fraction of the losses incurred by the IDF.

Realizing that Latrun would be a tough nut to crack, the IDF had begun searching for an alternative route to Jerusalem in order to bypass Latrun even before its second attempt to capture the area. By early June, such a route—nicknamed the "Burma Road" in honor of the famous Second World War supply line in Southeast Asia that bypassed Japanese army positions—had been not only found by IDF engineering forces, but also prepared by them for use.[21] The military and civilian supplies delivered via this route would sustain Jewish Jerusalem until the end of the War of Independence.

The creation of the Burma Road lessened the immediate strategic significance of Latrun to Israel. Still, both the Israeli government and the IDF remained determined to capture it; therefore, on June 9, IDF forces launched a third assault on the area, called Operation Yoram.[22] By this time, the battered 7th (IS) Infantry Brigade had been removed from the front line to rest and reorganize. It was replaced in the line by elements of the 10th (IS) and 11th (IS) Infantry Brigades. Under the cover of darkness and artillery fire, the 5th (IS) Infantry Battalion of the 10th (IS) Infantry Brigade and the 3rd (IS) Infantry Battalion of the 11th (IS) Infantry Brigade launched their assault. Due primarily to a navigational mishap, IDF forces were met by concentrated Arab Legion artillery, machine gun, and small arms fire. Though Israeli forces managed to break into Jordanian positions, the confused nature of the fighting spooked the IDF, which ordered its forces to retreat from the area before the fast-approaching daybreak, when they would be plainly visible to Arab Legion forces.

The last attempt to capture Latrun would occur in mid-July in the context of a much larger general offensive, Operation Dani, which was intended to expand Israeli territory in the narrow and vulnerable "waist" of the country, as well as to deliver a heavy blow to the Arab Legion.[23] While this general offensive turned out to be a considerable success, with the IDF capturing substantial tracts of territory in the vital and vulnerable center of Israel, as well as inflicting heavy losses on the Arab Legion, the final attempt to capture Latrun did not succeed, not least because the Jordanians had used the month since the IDF's third attempt to bolster their positions in and around Latrun, particularly by deploying a third infantry battalion to shore up the other two. From July 15–18, elements of the 10th and 11th (IS) Infantry Brigades, with elements of the 4th (IS) Infantry Brigade and the newly formed 8th (IS) Armored Brigade in support, sought to surround and overwhelm Arab Legion forces in the area. The pitched battles that accompanied the Israeli effort saw both sides engage in attacks and counterattacks against the other. IDF forces actually came within a hare's breath of capturing Latrun, but ultimately failed to do so when 8th (IS) Armored Brigade tanks, serving as part of a combined tank–infantry task force during a last-ditch attack, suddenly withdrew from the field as a consequence of a communications error, leaving the outgunned infantry with no choice but to retreat as well.[24]

All told, the IDF lost some 170 soldiers killed and about twice that number wounded in its abortive attempts to capture Latrun.[25] Matériel losses were considerable as well. Arab Legion losses in fighting men and fighting machines, though unknown with any precision, were significantly fewer than the losses incurred by the IDF. The Arab Legion, in short, had won an impressive victory over the IDF at Latrun. The IDF would have to wait 19 years, until the 1967 Six-Day War, to exact its revenge on the Arab Legion.

## Analysis of the Battle of Latrun

An analysis of the battle in light of the six variables follows.

### *Operational Planning*

While it improved somewhat over the course of the four attempts to capture Latrun, IDF operational planning overall during the battle can only be described as rather poor. First, instead of massing its available forces for concentrated assaults on the police fortress, the key position in the Arab Legion's entire defensive system at Latrun, the IDF repeatedly divided its forces in what turned out to be futile efforts to outflank and envelop all of the Jordanian forces in the area. It is pure speculation, of course, but the IDF might well have captured the fortress, causing the collapse of all Arab Legion forces in the Latrun area, had its forces launched a concentrated frontal assault against that position. Second, while the choice to launch night attacks made good sense in light of the strong Arab Legion defensive system around Latrun, these attacks began too late at night to give IDF forces sufficient time to accomplish their objectives, and the onset of daylight spelled disaster for those forces. And, third, the IDF's choices in respect of the forces designated to carry out the assaults at Latrun left much to be desired, particularly in regard to the first two assaults.

Operational planning was of less importance to Arab Legion forces, because they were on the defensive throughout the battle. Nevertheless, Jordanian operational planning proved to be first-rate. Not only did the Arab Legion consistently deploy its infantry battalions in well-chosen positions with clear fields of fire, but also these battalions were supported by well-sited AFVs and artillery, some of the latter mounted on the roof of the police fortress itself in order to take maximum advantage of their range and coverage in repelling IDF forces.

### *Intelligence and Reconnaissance*

Before the battle, the IDF did not possess accurate information on the nature and disposition of Arab Legion forces around Latrun. According to a prominent Israeli historian of the War of Independence, "The [IDF] went in blind; its eve-of-battle intelligence spoke of irregular 'local forces' rather than a[n Arab] Legion presence."[26] During the various stages of the battle itself, while Israeli reconnaissance patrols were apparently active in scouting the locations and deployments of Arab Legion forces, the information that they collected often seems not to have made its way up the IDF chain of command, rendering it useless to Israeli forces.[27] Over and over again, Arab Legion forces surprised IDF forces from previously unknown defensive positions. All in all, the absence of accurate and timely information about Arab Legion forces in the hands of the IDF before and during the battle contributed significantly to the Israeli defeat.

The Arab Legion, like the IDF, did not possess accurate information on the nature and disposition of Israeli forces on the eve of the battle. Nor does the Arab Legion appear to have conducted any reconnaissance patrols during the various stages of the battle. The dearth of information about its adversary's forces, however, did not hinder Arab Legion forces nearly to the same extent as it did IDF forces, as the former were content mainly to wage a defensive battle, which requires less in the way of accurate and timely intelligence and reconnaissance than an offensive battle.

### *Exploitation of Terrain*

The best use of terrain made by the IDF during the fighting at Latrun did not occur on the battlefield itself. Rather, it lay in the discovery and creation of the alternative supply route to Jewish Jerusalem that bypassed Arab Legion positions around Latrun. In respect of the battle, IDF forces repeatedly tried to exploit the hilly terrain in the Latrun area in order to hide themselves as they advanced to attack the flanks of Arab Legion positions; however, as a consequence either of night navigation errors or of a lack of information about the whereabouts of Arab Legion positions, they repeatedly failed to accomplish this task.[28] Israeli efforts to achieve tactical surprise through the exploitation of terrain for concealment, in other words, came to naught.

Arab Legion forces, to the contrary, made quite effective use of the hilly terrain around Latrun during the battle. They took advantage of their control of the high ground to create clear fields of fire below for their artillery, machine guns, and small arms. They also took advantage of the meandering terrain to hide many of their positions from the eyes of the IDF, giving Israeli forces a nasty shock on several occasions. Unquestionably, effective exploitation of terrain constituted a key ingredient in the Arab Legion victory at Latrun.

### *Quantity and Quality of Forces*

A standard rule of thumb among military experts holds that, all else equal, an attacking force must outnumber a defending force by at least a 3:1 ratio in order for the former to overcome the latter. IDF forces never enjoyed close to a 3:1 superiority over Arab Legion forces during any of the four attempts to capture Latrun. IDF forces, at most, may have enjoyed a slight numerical superiority in soldiers during the fighting. The Arab Legion, on the other hand, enjoyed a substantial numerical superiority in arms, especially artillery, throughout the battle. Numbers, in short, did not favor the IDF at Latrun.

Qualitatively speaking, some of the IDF forces deployed during the battle were simply not ready for combat—to say the least. The manpower of the 7th (IS) Infantry Brigade, for example, consisted primarily of raw recruits—new immigrants, many of them Holocaust survivors—who had received less than a week of training before being sent to the battlefield.[29] Moreover, these immigrants spoke at least eight different languages, rendering communication among them and with their officers problematic.[30] The quality of the brigade's junior officers also left much to be desired. The brigade's weapons were in equally bad shape. The other IDF brigades involved in the fighting had higher-quality and more cohesive manpower, but these brigades had been exhausted by prior battles, and they, too, were equipped mainly with low-quality weapons.[31]

The Arab Legion, organized and commanded by British officers under contract to the Hashemite Kingdom, is widely acknowledged to have been the best Arab fighting force in the War of Independence.[32] Its troops were well trained, especially in regard to waging defensive warfare. Its officer corps, including the Arab component of the corps, consisted of skillful and experienced professionals. And its British-supplied weapons were modern. The high quality of Arab Legion forces certainly played a considerable role in the Jordanian victory at Latrun.

### *Command and Control*

The entire IDF command and control ($C^2$) system at Latrun may be described as deficient—to put it politely. For starters, at the highest level of command, the headquarters (HQs) assigned to direct the four Israeli attempts to capture Latrun proved unable to coordinate the various forces in the field, so the individual IDF battalions involved in these assaults essentially operated in isolation from each other, allowing Arab Legion forces to repel them one at a time. Indeed, the commanding officer (CO) of the 7th (IS) Infantry Brigade—the brigade in charge of the first two assaults—was not even on scene, at least during the first assault, preferring instead to stay in the rear.[33] The HQ of the 11th (IS)

Infantry Brigade performed little better during the third assault.[34] And, during the fourth assault, a communications problem between HQ and forces in the field caused the latter's armor to withdraw from the battlefield just as these forces were on the verge of capturing the police fortress. Furthermore, IDF COs in the field sometimes lost control of their own soldiers, as demonstrated particularly by the unsanctioned, panic-stricken retreats that occurred during the first and second assaults. Not surprisingly, this phenomenon was especially evident in the ranks of the 7th (IS) Infantry Brigade, because its unseasoned, immigrant recruits could often not even understand their officers' orders.

The C² system of the Arab Legion, on the other hand, functioned quite smoothly throughout the battle. The firm discipline instilled in—as well as the extensive training of—its rank-and-file soldiers ensured that they faithfully carried out the orders of their officers and noncommissioned officers (NCOs). Even taking into account that C² is easier on the defensive than on the offensive, the Arab Legion's C² system indisputably outperformed the IDF's C² system at every level.

### *Combined Arms Warfare*

The IDF's performance in respect of combined arms warfare (CAW) during the Battle of Latrun can only be described as pitiful. During the first attempt to capture Latrun, IDF artillery and infantry forces acted completely independently of each other. Not only did the initial artillery barrage occur hours before the infantry forces even began their assault on Latrun, but also artillery fire was completely ineffectual in covering the infantry forces' retreat later on.[35] During the second attempt to capture Latrun, the IDF's practice of CAW showed some improvement, as artillery, mechanized infantry, and engineering forces worked together to penetrate the perimeter defenses of the police fortress; however, this attempt floundered when nonmechanized infantry forces failed to come to the assistance of the other IDF forces in order to exploit the breakthrough. And, the last attempt to capture Latrun fell apart when the armor in the armor–infantry task force suddenly abandoned the field after receiving an order to do so.

The Arab Legion, on the other hand, gets high marks for its practice of CAW during the battle. Its artillery and AFV forces closely supported its infantry forces throughout the fighting. "The Jordanians demonstrated a good grasp of combined-arms [sic] operations," according to a leading expert on Arab military performance during the

War of Independence, "regularly integrating infantry, armored cars, and artillery better than the Israelis."[36]

## Ramifications of the Battle of Latrun

By the time that the Battle of Latrun drew to a close, both Israel and the Hashemite Kingdom had reached a consensus that it was in their mutual interest to refrain from further large-scale hostilities against each other.[37] Thus, while both the IDF and Arab Legion would continue to conduct small-scale raids and engage in sniping against the other's forces for the duration of the War of Independence, they would not fight any further sweeping battles.

While Jewish Jerusalem survived to remain a part of Israel in spite of the IDF defeat at Latrun, the battle nevertheless had a couple of long-term negative strategic implications for the country. First, the Hashemite Kingdom's control of Latrun, aside from keeping the primary road to Jerusalem closed to Israeli traffic, represented a continuing threat to the integrity of the Jewish part of the city. It took the Six-Day War to remove the Sword of Damocles hanging over the head of Jewish Jerusalem. Second, the Arab Legion victory at Latrun helped to ensure that the Hashemite Kingdom emerged from the War of Independence in control of Judea and Samaria, meaning that Israel came out of the war with no strategic depth in the vital center of the country, where lay most of its population centers, industrial assets, and military installations. Indeed, at one point along its narrow waist, the country's width extended less than 10 miles (16 km). Not until after the Six-Day War, when Israel captured the whole of Judea and Samaria, would the country possess a measure of strategic depth along its central frontier.

# 2

# The First Battle of Abu Ageila/Um Katef

Despite incurring severe human, economic, and social costs, including the loss of one percent of its prewar population, the State of Israel emerged as the clear victor from its 1947–49 War of Independence.[1] Not only had the country inflicted far higher costs on its Arab adversaries, state and nonstate alike, but also it had acquired a significant amount of territory beyond the boundaries set by the 1947 United Nations (UN) partition plan, which had called for dividing western Palestine—that is, the approximately 20 percent of the territory originally allotted to the British-ruled Palestine Mandate that lay west of the Jordan River—into separate Jewish and Arab states.[2]

Nevertheless, Israel also emerged from its War of Independence in dire straits in respect of its national security, the country's survival far from assured in the future. After the guns fell silent, the Arab world continued to reject vehemently the very existence of a Jewish state, declaring itself fully committed to the eventual annihilation of that state. The combined effects of three variables in particular made this threat much more than simply an idle one.

First, the acquisition of additional territory during the War of Independence notwithstanding, Israel still had very unfavorable borders from a geographical perspective at the end of the hostilities. Not only were these borders extremely long in relation to the country's land area, but also they were quite flat, providing no topographical obstacles to invasion from the north, center, or south. Furthermore, all of the country's major urban areas, major industrial areas, and major military facilities lay not-too-distant from one or another of its borders. The center of the country—where resided most of the populace, industry, and military infrastructure—was very narrow, as little as under 10 miles (16 km) at one point. Israel, then, possessed neither defensible borders nor strategic depth in the wake of the War of Independence.

Second, numbers contributed another layer of uncertainty to the country's prospects for survival. With its small manpower base—the Jewish population upon independence consisted of a mere 600,000–650,000 souls—and a correspondingly small economic base, the Israel Defense Forces (IDF) could not dream of matching its Arab adversaries in terms of fighting men and fighting machines. Moreover, most of the manpower that Israel could spare for military purposes would of necessity consist of part-time reservists in order not to stunt the country's economic development. To try to match the Arab world fighting man for fighting man and fighting machine for fighting machine would invite socioeconomic collapse.

And, third, a lack of allies also contributed to the Israeli national security predicament. The rapidly mushrooming Cold War of the late 1940s–early 1950s induced both Western and Eastern bloc countries to seek the favor of the Arab world, which possessed seemingly unlimited stocks of oil, as well as militarily valuable territory. To befriend the Arab states, of course, Western and Eastern bloc countries had to remain at arm's length from Israel. This strategic reality ensured that the latter would not be welcome to join a bilateral or multilateral defense pact; therefore, not only could Israel not count on foreign assistance during a crisis, but also the country could even face foreign intervention on behalf of its adversaries during such a crisis. France and Great Britain, of course, would enter into an anti-Egyptian war pact with Israel on the eve of the 1956 Sinai Campaign; however, this union sprang from a unique set of circumstances that could not have been anticipated when Israel formulated its response to its national security dilemma in the early 1950s. Nor was a similar scenario likely to unfold again in the future.

By the early 1950s, the geographical, numerical, and diplomatic realities of the country's first years of existence convinced the Israeli defense establishment that the correct response to the country's national security predicament lay in an IDF war-fighting doctrine that emphasized offensive maneuver warfare. Because Israel lacked defensible borders and strategic depth, wars had to be fought on Arab territory. Hostilities on Israeli territory risked national destruction. Because Israel lacked numbers and allies, as well as economic staying power, wars had to be short and decisive. Long and indecisive hostilities risked allowing the Arab world to bring its much greater numbers into play, risked foreign intervention on the Arabs' behalf, and risked wrecking the Israeli economy. Offensive maneuver warfare, with its emphasis on swift and powerful thrusts deep into an adversary's rear in order to bring about the rapid collapse of its army, checked the right boxes in these respects. Moreover, offensive maneuver warfare would also permit the IDF to exploit to the fullest the considerable qualitative advantage that its manpower enjoyed vis-à-vis the manpower of its Arab adversaries—offensive maneuver warfare, put another way, favored the higher-quality "few" against the lower-quality "many."

While the Israeli defense establishment was united with respect to the notion that the IDF must place offensive maneuver warfare at the heart of its war-fighting doctrine, the defense establishment was not united with respect to the appropriate force structure that the IDF ought to adopt in order to wage offensive maneuver warfare.[3] On the one hand, a small "tank school" promoted the idea that IDF armored columns should both punch through an adversary's defenses and then exploit the breakthroughs to drive into that adversary's rear, with airpower in close support of the advance of these columns. Infantry, artillery, and engineers would play secondary roles within this paradigm of offensive maneuver warfare, mopping up the isolated and disorganized remnants of an adversary's forces that had been left behind by the advancing armored columns.

The IDF's war-fighting doctrine, especially after the 1967 Six-Day War, would rather firmly embrace this version of offensive maneuver warfare; however, prior to the Sinai Campaign, a large "mobile infantry school" held the upper hand in the debate over what version of offensive maneuver warfare ought to be adopted by the IDF. This school promoted the idea that mechanized infantry should take the lead role in piercing an adversary's defenses and then breaking out into its rear. Armor, artillery, and engineers would play secondary roles within this paradigm of offensive maneuver warfare. The fact that the IDF Armored Corps experienced repeated and major problems during the early 1950s in keeping its tanks serviceable ultimately swung the debate in favor of the mobile infantry school. Indeed, during one field exercise held by the IDF just a few months prior to the Sinai Campaign, about two thirds of the tanks involved in the maneuvers suffered mechanical breakdowns.[4] Mechanized infantry, in the eyes of the Israeli defense establishment, appeared to be considerably more reliable, more flexible, and more mobile than armor.

## Prelude to the First Battle of Abu Ageila / Um Katef

Though Israel and Egypt had signed an armistice agreement at the end of the War of Independence, quiet along the Israel–Egypt border did not persist for very long thereafter. Armed Palestinian infiltrators from Gaza—a territory that had remained under Egyptian occupation following the war—eventually began to wreak havoc on Israeli frontier communities. Deadly incursions across the border, in turn, elicited IDF retaliatory raids against Gaza, which escalated into large-scale engagements between Israeli and Egyptian forces by the mid-1950s.[5] The ongoing Egyptian blockade of the Straits of Tiran, which prevented Israel from engaging in maritime commerce via the Red Sea, also stoked tensions between the two countries.[6]

Nevertheless, neither border skirmishing nor maritime blockade rendered inevitable a second round of war between Israel and Egypt. Rather, the event that would tip the scales in favor of war occurred in 1955.[7] During the middle of that year, employing Czechoslovakia as an intermediary, the Soviet Union agreed to supply Egypt with a massive stockpile of arms, including fighter-bombers, tanks, armored personnel carriers (APCs), artillery, and submarines.[8] The addition of hundreds of modern fighter-bombers, tanks, APCs, and guns to the Egyptian arsenal would seriously undermine in both a quantitative and a qualitative sense the balance of power between the IDF and the Egyptian army.

Consequently, the "Czech" arms deal, as it came to be known, instantly set off alarm bells within the Israeli defense establishment. If the Egyptian army were to be permitted to absorb the new arms into its arsenal, reckoned the defense establishment, Israel would become vulnerable to invasion and destruction. How best, then, to ensure that Egypt would not reap the benefits of this arms deal?[9] One school of thought within the defense establishment—the minority school led by the IDF chief of staff—argued for an immediate preventive war against Egypt, before the arms could be taken into inventory by the Egyptian army. But another school of thought—the majority school led by the prime minister—did not want to initiate a war with Egypt until the country had first secured a great power ally for such an endeavor.

The opportunity to acquire a great power ally—two, in fact—arose in the middle of 1956. Both France and Great Britain had reasons of their own for going to war against Egypt.[10] Both countries opposed strenuously the Egyptian government's decision to nationalize the Suez Canal Company. The French government also objected vehemently to the Egyptian government's not-so-covert support of nationalist forces seeking to oust France from Algeria, while the British government objected strongly to the Egyptian government's equally not-so-covert support of nationalist forces seeking to overthrow pro-Western monarchies in countries like Jordan and Saudi Arabia.

Neither country, however, felt that it could mount an unprovoked invasion of Egypt, especially in view of the refusal of the United States to countenance such a course of action. By the autumn of 1956, therefore, France (enthusiastically) and Great Britain (much less so) sought to bring Israel into the equation in order to justify a war against Egypt. Following intense—and often difficult—negotiations among the three governments, to make a long story short, they hammered out a mutually acceptable plan to invade Egypt.[11] Codified in the Sèvres Protocol, the plan called for the IDF to conduct an operation in the vicinity of the Suez Canal—a parachute infantry drop near the Mitla Pass in the Sinai—in order to provoke a clash between the IDF and the Egyptian army. At this juncture, the French and British governments would demand an immediate cessation of hostilities, as well as a withdrawal of both IDF and Egyptian army forces from the canal zone. Ostensibly in order to guarantee "freedom of passage" through the Suez Canal, French and British forces would then occupy the area. The Israeli government, of course, would agree to the Franco–British demand. When the Egyptian government refused to do so—and all three governments knew that it could not do otherwise—Israel would have a pretext for capturing all of the Sinai and Gaza, while France and Great Britain would have a pretext for seizing the Suez Canal. All three governments hoped that, as a result of these operations, the Egyptian government would collapse, to be replaced by a more moderate one.

On the eve of the Sinai Campaign, the IDF had assembled a total of 45,000 men in 10 brigades for the upcoming offensive into the Sinai and Gaza: nine mechanized and nonmechanized infantry brigades, including one parachute infantry brigade, and one armored brigade.[12] Of the approximately 200 tanks assembled for the campaign, about two thirds had been parceled out among the infantry brigades, with the rest concentrated in the armored brigade. For its part, the Egyptian army had two infantry divisions, as well as assorted independent forces, with a total of about 30,000 men and approximately 200 tanks deployed throughout the Sinai. Most of the Egyptian army's order of battle (ORBAT) remained west of the Suez Canal in order to counter a potential Franco–British invasion of the canal zone. What Egyptian army forces in Sinai and Gaza lacked in numbers, however, they made up for in terms of occupying heavily fortified defensive positions.

Operation Kadesh 2, the codename for the IDF offensive, commenced during the late afternoon of October 29, 1956, when Israel Air Force (IAF) transport aircraft dropped the 890th (IS) Parachute Infantry Battalion of the 202nd (IS) Parachute Infantry Brigade near the Mitla Pass.[13] Over the next few days, IDF and Egyptian army forces would fight four major engagements: at the aforementioned Mitla Pass (in the central Sinai), at Rafah (in northeastern Sinai/Gaza); at Sharm el-Sheikh (at the southern tip of the Sinai); and at Abu Ageila/Um Katef (in northeastern Sinai).

## The First Battle of Abu Ageila/Um Katef

During the final stage of the War of Independence, the IDF launched a brief and shallow incursion into the Sinai as part of its operational plan to deliver a coup de grâce to the Egyptian army. The Abu Ageila/Um Katef defensive complex fell to Israeli forces during the incursion.[14] The "battle" that resulted in the capture of the complex, however, could not really be described as anything of the sort, as the small Egyptian garrison stationed there essentially surrendered to equally small Israeli forces without putting up much of a fight. Hence, the first genuine battle between the IDF and the Egyptian army at the Abu Ageila/Um Katef complex would not occur until the Sinai Campaign.[15]

By the time that the IDF launched Kadesh 2, the Egyptian army had transformed what had formerly been a rather minor outpost into its most heavily fortified position in the Sinai. The 6th (EG) Infantry Brigade of the 3rd (EG) Infantry Division, comprised of three infantry battalions, constituted the principal formation at the complex. Self-propelled antitank guns (in the form of 20–30 tank destroyers), as well as artillery, antiaircraft, and reconnaissance forces, had been attached to the brigade.[16] The Egyptian army had also deployed another infantry battalion at Quseima, a separate defensive complex located a few miles southeast of the Abu Ageila/Um Katef complex, close enough to lend a hand in the latter's defense. While the 6th (EG) Infantry Brigade had no tanks of its own, armored forces stationed in the central Sinai could be called upon to come to its assistance in case of need. For its part, IDF general headquarters (GHQ) assigned the task of capturing the Abu Ageila/Um Katef complex to the 38th (IS) Division, an all-arms division, whose ORBAT consisted of the 4th and 10th (IS) Infantry Brigades, the 7th (IS) Armored Brigade, and attached artillery, engineering, and reconnaissance forces. The 38th (IS) Division had as many as 100 tanks at its disposal at the outset of the battle, but many of these vehicles would be diverted to other tasks in the event and, therefore, would not see action at Abu Ageila/Um Katef.

The fighting at Abu Ageila/Um Katef commenced during the morning of October 30, when the 4th (IS) Infantry Brigade crossed the border headed for Quseima, followed soon thereafter by the 7th (IS) Armored Brigade. After an exchange of fire between Israeli and Egyptian forces, in which IDF tank fire proved decisive, the Egyptian garrison at Quseima simply withdrew to the Abu Ageila/Um Katef fortifications, though some of the soldiers fled into the desert, while still others were taken captive by Israeli forces.[17]

IDF GHQ, upon consulting with IDF Southern Command headquarters (HQ) and the 38th (IS) Division commander later in the morning, then diverted the 4th (IS) Infantry Brigade southward, instructing it to bypass the Abu Ageila/Um Katef complex, in order to screen and assist elements of the 202nd (IS) Parachute Infantry Brigade, which had begun an advance into central Sinai to reach the Mitla Pass in order to reinforce the 890th (IS) Parachute Infantry Battalion. The participation of the 4th (IS) Infantry Brigade in the battle essentially came to an end at this point, as did the participation of much of the 7th (IS) Armored Brigade, elements of which were also instructed by GHQ to bypass the Abu Ageila/Um Katef complex in order to assist the advance of the 202nd (IS) Parachute Infantry Brigade, as well as in order to be in a position to block Egyptian armor stationed in the central Sinai from coming to the aid of the Egyptian garrison at Abu Ageila/Um Katef.[18]

Concomitantly, GHQ ordered the 82nd (IS) Armored Battalion of the 7th (IS) Armored Brigade to assist in the capture of the Abu Ageila/Um Katef complex by outflanking the fortifications from the south and then attacking from the west. Negotiating its way over very difficult terrain, the 82nd (IS) Armored Battalion quickly captured Abu Ageila itself, essentially cutting off the complex from the rest of the Egyptian army's forces in the Sinai. While the 82nd (IS) Armored Battalion dug in and prepared for an assault on the Ruafa Dam, a key strongpoint within the Abu Ageila/Um Katef complex, other elements of the 7th (IS) Armored Brigade, in tandem with interdiction strikes by IAF fighter-bombers, blocked Egyptian reinforcements, especially the 1st (EG) Armored Brigade of the 4th (EG) Armored Division, from coming to the aid of the Egyptian garrison at the complex.[19]

With the 82nd (IS) Armored Battalion now in position to launch an attack from the west on the Ruafa Dam, IDF GHQ moved the 10th (IS) Infantry Brigade into position to launch an attack on Um Katef itself from the east. What was supposed to be a coordinated "two-front" assault—perhaps even a "three-front" assault if some stray forces belonging to the 7th (IS) Armored Brigade to the south of the Abu Ageila/Um Katef complex are counted in the mix—turned out to be nothing of the sort and bogged down very quickly. The Egyptian garrison easily beat back the initial 10th (IS) Infantry Brigade attack, which took place during the afternoon of October 31. The garrison also beat back a more determined second assault by the 10th (IS) Infantry Brigade during the night of October 31–November 1. During the morning of November 1, IDF forces mounted yet a third assault—this time around carried out by elements of the 37th (IS) Mechanized Infantry Brigade (a unit under the control of IDF GHQ rather than the 38th (IS) Division)—against Um Katef's defenses. Resulting only in significant casualties to Israeli forces, this assault, too, did not breach Um Katef's defenses.[20]

Only on the western side of the Abu Ageila/Um Katef complex did the attacking IDF forces meet with success. Even at the Ruafa Dam, however, Egyptian forces put up unexpectedly stiff resistance at first, stalling the 82nd (IS) Armored Battalion's initial assault. But the battalion regrouped and eventually overcame Egyptian forces there after a difficult and costly fight that lasted well into the night of October 31.[21]

The IDF finally "captured" the entire Abu Ageila/Um Katef defensive complex on November 2, after Egyptian army GHQ had instructed all of its Sinai-based forces to withdraw toward the Suez Canal to avoid the possibility of being cut off from the rest of the army in Egypt by a Franco–British invasion and occupation of the Suez Canal zone.[22] Slipping through the Israeli noose around Abu Ageila/Um Katef, the garrison retreated toward El-Arish and, thence, back to Egypt. Only part of the garrison completed the trek, however. IDF forces captured hundreds of Egyptian soldiers along the route of their sojourn, while an undetermined number of other soldiers got lost and perished in the desert.[23]

The battle itself essentially ended in a draw. While IDF forces had managed to take Quseima, Abu Ageila, and the Ruafa Dam, as well as to surround the entire defensive complex, they had not been able to take Um Katef, the core of the complex. While the Egyptian garrison had put up a stout defense at Um Katef for more than two days, had the garrison not withdrawn when it did so, this position, too, would most likely have fallen to IDF forces shortly thereafter, because not only were the defenders surrounded, but also they were almost out of water and ammunition.

In respect of losses in fighting men and fighting machines, all that can be said with certainty is that the Egyptian army lost more men and machines at Quseima, Abu Ageila, and the Ruafa Dam than IDF forces. Conversely, IDF forces lost more fighting men and fighting machines at Um Katef than the Egyptian army. All told, Egyptian army losses in men were higher than IDF losses, particularly if the Egyptian soldiers who perished in the desert are added to the total figure. Likewise, Egyptian army losses in machines were higher, especially in light of the fact that the 6th (EG) Infantry Brigade abandoned its equipment when it withdrew from Um Katef.

## Analysis of the First Battle of Abu Ageila / Um Katef

An analysis of the battle in light of the six variables follows.

### *Operational Planning*

The rapidity and secrecy with which the French, British, and Israeli governments hammered out and implemented their war plan against Egypt impeded IDF operational planning for the battle. "Piecemeal and tentative," according to the leading American military historian of the battle, rather than systematic and deliberate, best describe the nature of that planning.[24] First, the overriding concern to preserve secrecy meant that, while IDF GHQ participated in the formulation of the joint war plan, both Southern Command HQ and 38th (IS) Division command had to be kept in the dark, creating in turn an atmosphere of confusion and tension between GHQ and its subordinate commands. Neither Southern Command HQ nor 38th (IS) Division command, for example, was aware of the diplomatic rationale behind what they considered to be the flawed prewar GHQ decision to withhold the 7th (IS) Armored Brigade from battle during the first days of the war.[25] Second, the need to assist the exposed 890th (IS) Parachute Infantry Battalion, which was to be dropped deep in the Sinai in order to serve as a political pretext for the Franco–British invasion of the Suez Canal zone rather than on grounds of military logic, meant that much of the 38th (IS) Division—namely, the 4th (IS) Infantry Brigade and much of the 7th (IS) Armored Brigade—would be diverted away from the battle, violating the crucial principle of concentration of forces in the attack.[26] And, third, the IDF assigned the most difficult task in the battle—the capture of Um Katef itself—to the 10th (IS) Infantry Brigade, a unit that simply lacked the capability to achieve this objective.[27] Israeli operational planning, then, did not bode well insofar as concerned the IDF's prospects for victory in the upcoming battle.

The Egyptian army did not anticipate either the scope or the timing of the Israeli invasion of the Sinai. Despite the lack of a coherent operational plan to counter the IDF's moves, however, Egyptian army GHQ responded aggressively to the Israeli invasion by dispatching reinforcements to threatened positions, though these reinforcements could be rather sluggish in their movements to the front.[28] For its part, the 6th (EG) Infantry Brigade competently deployed its forces in and around the Abu Ageila/Um Katef complex in order to meet an Israeli attack, though it neglected to prepare adequately for the possibility that IDF forces might outflank the complex from the south and west.[29]

### *Intelligence and Reconnaissance*

Neither the IDF nor the Egyptian army distinguished itself in respect of intelligence collection prior to the battle or reconnaissance probes during the battle. The IDF underestimated the Egyptian army's capabilities before the war, especially when the latter's forces had an opportunity to fight from well-prepared defensive positions. Indeed, the IDF chief of staff's attitude toward the Egyptian army on the eve of the Sinai Campaign could best be described as dismissive.[30] Moreover, the IDF possessed very limited information about the layout of the 6th (EG) Infantry Brigade forces at Abu Ageila/Um Katef.[31] Insufficient reconnaissance probes during the battle itself contributed to the debacles suffered by the 10th (IS) Infantry Brigade and 37th (IS) Mechanized Infantry Brigade. The lone triumph in regard to IDF intelligence and reconnaissance efforts—and it proved to be a very significant triumph—was registered by the 82nd (IS) Armored Battalion's reconnaissance force, which discovered the route used by the battalion to outflank the Abu Ageila/Um Katef complex from the south and west.[32]

Like the IDF before the Sinai Campaign, the Egyptian army underestimated its adversary's capabilities. Unlike the IDF, however, the Egyptian army prior to the war lacked the means to gather high-quality intelligence about its adversary, so it might be forgiven for this blunder. Less understandable, however, is the 6th (EG) Infantry Brigade's apparent unwillingness to conduct reconnaissance probes during the battle in order to determine the precise whereabouts of its adversary, especially after the fall of outlying observation posts had nullified the advantage provided to it by the presence of forward observers.

### *Exploitation of Terrain*

The natural terrain in and around the Abu Ageila/Um Katef complex strongly favored the defender.[33] The complex itself was built atop a plateau and "guarded" by hills along its southern flank and sand dunes along its northern flank, leaving no easy approach for an attacker from these directions. Moreover, several prominent hills to the east and northeast of Abu Ageila/Um Katef screened the complex from these directions, again leaving no easy approach for an attacker.

To enhance the defensibility of this already formidable natural terrain, the Egyptian army added an intricate and dense series of bunkers, trenches, forward observation posts, minefields, barbed-wire fences, antitank obstacles, and pre-sighted artillery "killing grounds."[34] This system of mutually supportive positions would play a conspicuous role in the debacles suffered by the 10th (IS) Infantry Brigade and 37th (IS) Mechanized Infantry Brigade in their frontal assaults against Um Katef. On the other hand, the 6th (EG) Infantry Brigade neglected to enhance adequately the defensibility of the southern flank of the Abu Ageila/Um Katef complex, thinking that tanks and other large armored vehicles could not possibly negotiate the terrain there. This oversight would cost the brigade in the upcoming battle.

For its part, the 7th (IS) Armored Brigade exploited the terrain on the southern side of the Abu Ageila/Um Katef complex brilliantly in order to outflank the 6th (EG) Infantry Brigade and to attack the latter's positions from the much more sparsely defended western side of the complex. This very profitable instance of terrain exploitation makes it all the more mystifying as to why other elements of the 38th (IS) Division, namely the 10th (IS) Infantry Brigade and the 37th (IS) Mechanized Infantry Brigade, made no efforts of their own at terrain exploitation.

## *Quantity and Quality of Forces*

Quantitatively speaking, IDF forces had a considerable advantage over Egyptian army forces in both fighting men and fighting machines on the eve of the battle.[35] The 38th (IS) Division—not even counting the elements of the 37th (IS) Mechanized Infantry Brigade that joined the battle later—boasted approximately 12,000 soldiers, as many as 100 tanks, and ample quantities of APCs, guns, and other arms. The 6th (EG) Infantry Brigade, including the infantry battalion initially stationed at Quseima, in contrast, mustered approximately 3,500 soldiers, 20–30 tank destroyers (but no tanks), and smaller numbers of APCs, guns, and other arms than its adversary.

The raw numbers, however, are misleading here. While all of the 6th (EG) Infantry Brigade's fighting men and fighting machines participated in the battle, not all of the 38th (IS) Division's men and machines did so. The 4th (IS) Infantry Brigade and much of the 7th (IS) Armored Brigade departed from the battle after the brief engagement at Quseima in order to undertake other tasks not connected to the capture of the Abu Ageila/Um Katef complex. Furthermore, Israeli command and control ($C^2$) problems throughout the battle were such that the remaining attackers were never able to bring their superior numbers to bear against the defenders. Indeed, it is not at all clear that IDF forces even outnumbered Egyptian army forces in fighting men and fighting machines during the engagements at Abu Ageila, the Ruafa Dam, and Um Katef. The former certainly did not ever come close to enjoying the 3:1 advantage in men and machines that has traditionally been deemed necessary for an attacker to overcome a defender.

Qualitatively speaking, in respect of fighting machines, both IDF and Egyptian army forces were equipped mainly with a hodgepodge of Second World War-era tanks, tank destroyers, guns, APCs, and other arms, some of them upgraded, in particular by the IDF Ordnance Corps.[36] Israeli forces were equipped primarily with American- and French-produced machines, while Egyptian forces were equipped primarily with British-produced machines, the fruits of the recent Soviet–Egyptian arms deal not yet having filtered down to the 6th (EG) Infantry Brigade. All in all, neither the IDF nor the Egyptian army possessed a marked advantage in fighting machines.

Much the same could be said of the manpower involved in the battle. The officers and men of the 7th (IS) Armored Brigade proved to be of first-rate quality not only at Abu Ageila/Um Katef, but also throughout the rest of the Sinai Campaign. Elements of this brigade pulverized much of the 1st (EG) Armored Brigade in armor-versus-armor clashes in central Sinai toward the end of the war.[37] The same could not be said about the officers and men of the 10th (IS) Infantry Brigade or the 37th (IS) Mechanized Infantry Brigade. The former, commanded by an artillery officer not versed in infantry tactics, was composed of older reservists who had not received any training as assault infantry.[38] The latter, commanded by an overzealous officer who launched a foolhardy frontal attack in broad daylight against Um Katef, had not received adequate amounts of manpower, equipment, or training before the Sinai Campaign.[39]

Like their Israeli counterparts, Egyptian forces at Abu Ageila/Um Katef presented a rather mixed picture. Some officers and men of the 6th (EG) Infantry Brigade gave quite a good account of themselves, especially at Um Katef and at the Ruafa Dam. Other officers and men, to the contrary, did not fight nearly as well, especially at Quseima and Abu Ageila.[40]

## *Command and Control*

IDF forces experienced $C^2$ problems during the battle at Abu Ageila/Um Katef.[41] The problems actually began on the eve of the war. Not privy to the particulars of the Sèvres Protocol, neither IDF Southern Command HQ nor 38th (IS) Division command understood the diplomatic reason behind IDF GHQ's refusal to commit the 7th (IS) Armored Brigade to the fight immediately upon the outbreak of war. To IDF Southern Command and the 38th (IS) Division, this refusal seemed to run counter to military logic. On the first day of the war, therefore, IDF Southern Command HQ defied IDF GHQ's instructions and sent the 7th (IS) Armored Brigade into battle. While certainly justifiable from a purely military perspective, this decision might have had extremely harmful consequences for Israel had events unfolded differently.

On this occasion, IDF Southern Command HQ's decision ultimately worked out for the best. Indeed, IDF GHQ later approved of the decision, with the IDF chief of staff saying that, "Better to be engaged in restraining the noble stallion than in prodding the reluctant mule!"[42] Still, IDF Southern Command HQ's decision was symptomatic of the tendency throughout the battle to ignore the chain of command—a tendency that caused serious $C^2$ problems on other occasions during the battle. IDF GHQ and Southern Command HQ routinely bypassed the 38th (IS) Division command in order to issue instructions directly to the 7th (IS) Armored Brigade, 10th (IS) Infantry Brigade, and 37th (IS) Mechanized Infantry Brigade—instructions that were not only often contradictory in nature, but also consistently undermined the ability of the 38th (IS) Division command to direct the battle. The confusion and tension created by this splintered $C^2$ system clearly impeded the ability of the 7th (IS) Armored Brigade, 10th (IS) Infantry Brigade, and 37th (IS) Mechanized Infantry Brigade to coordinate their assaults on Abu Ageila/Um Katef. $C^2$ problems also indisputably contributed to the bungled assaults on Um Katef by the 10th (IS) Infantry Brigade and the 37th (IS) Mechanized Infantry Brigade, as well as a tragic friendly fire incident between elements of the 37th (IS) Mechanized Infantry Brigade and elements of the 82nd (IS) Armored Battalion at the end of the battle that led to losses in fighting men and fighting machines.[43]

Moreover, some IDF officers were simply not up to the tasks assigned to them. The 10th (IS) Infantry Brigade commander had no idea about how to mount an infantry assault, and he was relieved of command after two bungled efforts to do so. The 37th (IS) Mechanized

Infantry Brigade commander did not sufficiently prepare for his unit's assault, and both he and his men paid with their lives in an attack whose sole result was to add to the roster of IDF losses. Only the officers of the 7th (IS) Armored Brigade displayed a consistently high level of command ability during the battle for Abu Ageila/Um Katef, but even they sometimes encountered $C^2$ problems as a consequence of having to rely on inadequate radio equipment for communications.[44]

Egyptian army forces at Abu Ageila/Um Katef had fewer problems than IDF forces in respect of $C^2$, which is not surprising in view of the fact that defenders frequently experience fewer problems in this regard than attackers.[45] That said, the 3rd (EG) Infantry Division HQ gave considerable latitude to the 6th (EG) Infantry Brigade to conduct the defense of Abu Ageila/Um Katef as the latter saw fit. Furthermore, the 6th (EG) Infantry Brigade was blessed with a solid cadre of officers and noncommissioned officers (NCOs) who were well versed in the particulars of waging a defensive battle from fortified positions. Nevertheless, the 6th (EG) Infantry Brigade was certainly not without $C^2$ problems of its own, as demonstrated by its very poor performance during the engagements at Quseima and Abu Ageila.

### *Combined Arms Warfare*

IDF forces at Abu Ageila/Um Katef, on the whole, did not distinguish themselves in respect of the practice of combined arms warfare (CAW). Elements of the 7th (IS) Armored Brigade and the 4th (IS) Infantry Brigade formed an effective armor–infantry team at Quseima, while the 82nd (IS) Armored Battalion productively integrated armor, mechanized infantry, and engineers to outflank Abu Ageila/Um Katef, as well as to capture Abu Ageila and, later, the Ruafa Dam. Apart from these episodes, however, IDF forces essentially neglected the practice of CAW during the battle. The 10th (IS) Infantry Brigade and the 37th (IS) Mechanized Infantry Brigade launched abortive frontal assaults at Um Katef with nonmechanized and mechanized infantry, respectively, without any armor support and with little in the way of artillery or engineering support. For their part, IAF fighter-bombers carried out approximately 50 sorties against the Abu Ageila/Um Katef defensive complex during the battle, but apparently none of them in close coordination with ground forces, so they had only a minimal impact on the course of the fighting.[46] The IAF's only tangible contribution in support of the ground forces occurred when its fighter-bombers helped to scatter and turn back Egyptian army reinforcements on their way to assist their compatriots at Abu Ageila/Um Katef.[47]

The 6th (EG) Infantry Brigade's efforts to engage in the practice of CAW during the battle were also uneven. On the one hand, Egyptian forces productively integrated infantry, artillery, and tank destroyers in their steadfast defense of Um Katef, as well as during the early hours of the fight for the Ruafa Dam. On the other hand, during the final hours of the fight for the Ruafa Dam, as well as during the engagements at Quseima and Abu Ageila, the practice of CAW was not in evidence. For its part, the Egyptian air force made no effort to assist Egyptian ground forces at Abu Ageila / Um Katef, even before French and British airpower destroyed most of its aircraft at the outset of the Franco–British invasion of the Suez Canal zone.

## Ramifications of the First Battle of Abu Ageila / Um Katef

The Sinai Campaign came to an end on November 5. The IDF had won an impressive victory over the Egyptian army. Not only had Israeli forces captured all of the Sinai, but also they had done so at modest cost to themselves. The IDF lost approximately 200 soldiers and 25 tanks.[48] The Egyptian army experienced much higher losses—approximately 1,000 soldiers and 100 tanks, perhaps more. Egyptian forces also abandoned a substantial amount of equipment in the field, including tanks and guns, some of which the IDF later incorporated into its ORBAT.

The battle at Abu Ageila/Um Katef, however, constituted an anomaly, ending in a draw, the only major battle of the Sinai Campaign that the IDF did not win outright. Be that as it may, the outcome of the battle had no tangible effect on the overall direction taken by the Sinai Campaign. The stout defense put up by Egyptian forces simply did not prevent Israeli forces from bypassing the Abu Ageila/Um Katef complex in their drive to penetrate into the interior of the Sinai on their way to the Suez Canal.

In the wake of the Sinai Campaign, to its credit, the IDF carefully studied its less-than-stellar performance during the First Battle at Abu Ageila / Um Katef, drawing lessons from its encounter there that would be applied with devastating effect in its next encounter with the Egyptian army at the complex during the Six-Day War.

# 3

# The Second Battle of Abu Ageila/Um Katef

The Israel Defense Forces (IDF) registered an impressive victory in the 1956 Sinai Campaign, routing the Egyptian army in the Sinai and conquering the entire peninsula in just a few days, thereby validating its prewar decision to adopt offensive maneuver warfare as the basis of its method of fighting an interstate war. Nevertheless, the lessons derived by the IDF in the aftermath of the hostilities prompted it to modify its approach to offensive maneuver warfare. Whereas mechanized infantry had been the leading arm during the war, with armor and airpower acting in supporting roles, their indispensable contributions to the IDF's triumph in the fighting convinced the Israeli defense establishment that armor and airpower should be placed at the center of the IDF's approach to offensive maneuver warfare moving forward. A prominent American military historian of the Arab–Israeli conflict remarked that:

> The 1956 campaign in the Sinai had demonstrated to the senior leadership of the army the critical importance of air power [sic] in modern warfare: control of the air had tremendously enhanced Israeli ground operations in the desert.[1]

He went on to observe that:

> Instead of an infantry-dominated army, military planners now envisioned a rapidly advancing army led by large tank formations that employed great mobility and firepower. This force's objective would be to strike into the enemy's rear, causing its army to collapse through the disruption of its command and control system and lines of communication. Such thinking spelled the end

> of the doctrinal dispute generated by the [pre-Sinai Campaign] armor-versus-infantry debate.[2]

The switch from an emphasis on mechanized infantry to an emphasis on armor and airpower had implications for the IDF's force structure and war-fighting doctrine.

In respect of force structure, the IDF devoted the lion's share of its resources between the Sinai Campaign and the 1967 Six-Day War to the growth of its air and armored forces. The fighter-bomber inventory of the Israel Air Force (IAF) ballooned by more than 50 percent during this decade. Furthermore, the air force replaced all of its obsolescent piston-engined aircraft, which had constituted more than 50 percent of its inventory during the Sinai campaign, with modern French jet aircraft. This development, coupled with the acquisition of more and better ordnance, increased greatly the IAF's range and striking power.[3] The IDF armored corps expanded at an even more prodigious pace. The number of tanks in its inventory tripled between 1956 and 1967. Moreover, whereas it entered the Sinai Campaign with mostly obsolescent tanks—Second World War-vintage machines that had been acquired from the scrap heaps of Western armies—it entered the Six-Day War with a mix of modern, post-Second World War British and American tanks, as well as with upgraded Second World War-vintage fighting machines that could now slug it out with the latest Soviet and American machines in Arab hands.[4]

The rapid growth of the IAF and IDF armored corps left rather few resources to invest in the development of other arms. The mechanized infantry entered the Six-Day War riding atop the same Second World War-vintage half-tracks that it had employed during the Sinai campaign, the IDF not making any effort to obtain modern armored personnel carriers (APCs) during the interwar years.[5] Similarly, though a substantial part of the artillery corps fought with self-propelled guns during the Six-Day War, a development in tune with the IDF's reliance on offensive maneuver warfare, these guns were usually not available in significant numbers, nor were they generally of the most modern types. The infantry, both mechanized and nonmechanized (including parachute infantry), went into battle with small arms and antitank weapons of inferior quality in comparison to the small arms and antitank weapons in Arab hands. Engineering forces possessed insufficient and outdated equipment. And, finally, the logistics corps' heavy reliance on civilian trucks, buses, and even automobiles to move fighting men and fighting machines to, around, and from the battlefields would have been inconceivable in Arab armies, let alone Western- and Soviet-bloc armies.

In respect of war-fighting doctrine, the IDF's approach to offensive maneuver warfare between the Sinai Campaign and the Six-Day War stressed the importance of obtaining air superiority immediately upon the outbreak war.[6] To this end, IDF general headquarters (GHQ) assigned the task of destroying the adversary's airpower through a massive attack on its bases at the outset of hostilities to the IAF. Simultaneously, IDF armored forces would punch through the adversary's frontline forces, breaking out into their rear areas. Once air superiority had been established over the battlefield and armored breakthroughs effected on the battlefield, airpower and armor would then team up to rout the (by now presumably disoriented and demoralized) adversary's remaining forces in a series of swift-moving thrusts.

Be that as it may, the extent to which the IDF's pre-1967 war-fighting doctrine focused on the air–armor combination to the exclusion of other arms should not be overdone.[7] Before the Six-Day War, the IDF's doctrine also stressed the importance of interarms cooperation within the ground forces at both the tactical and operational levels. The plans for the break-in battles on the Sinai front—the only battles planned before the war—envisioned all-arms divisions composed of tanks, mechanized and nonmechanized infantry, artillery, and engineers engaging in traditional combined arms warfare (CAW) in order to pierce Egyptian army defenses.[8] Indeed, of the seven divisions sent into battle by the IDF during the Six-Day War—three on the Sinai front, two on the Judean and Samarian front, and two on the Golan front—no less than six were more or less balanced all-arms divisions.[9] The IDF deployed just one "armor-only" division—on the Sinai front—because the terrain over which it had to pass was judged to be unsuitable for vehicles other than tanks in light of the projected speed of its advance.[10]

## Prelude to the Second Battle of Abu Ageila / Um Katef

The origins of the Six-Day War lay in border skirmishes between Israel and its Arab neighbors, particularly Syria.[11] In the spring of 1967, tensions between Israelis and Arabs began to escalate toward war after IAF fighter-bombers shot down six of their Syrian counterparts in an air battle. Always keen to stir the pot in the Middle East, the Soviet Union took advantage of the air battle to spread a false accusation to the effect that Israel was about to invade Syria. In response to the Soviet accusation, Egypt—the leader of the Arab world—demanded the removal of United Nations (UN) peacekeepers from the Sinai, mobilized its army, and deployed much of it to the peninsula in order to counter "Israeli aggression." Israel, predictably, mobilized the IDF in order to counter Egyptian military moves. Egypt also blockaded the Straits of Tiran, cutting off Israeli access to the Red Sea, and Jordan entered into an anti-Israel war pact with Egypt and Syria. Iraq and other Arab countries deployed forces in support of the frontline Arab states. Meanwhile, an American effort to defuse the latest Arab–Israeli crisis through diplomacy—an effort that had the endorsement of Israel—ultimately came to naught as a result of Arab and Soviet intransigence, as well as Western and Third World indifference. Upon the collapse of this American effort, the Israeli government concluded that its only remaining option was to embark on war, as the state's very survival now seemed to be at risk.

The IDF based its post-1956 war-fighting doctrine on the assumption that Israel would deliver the first blow in the next round of hostilities. On the eve of the Six-Day War, it deployed its ground forces on the Sinai front in a manner that reflected this assumption. The IDF grouped the bulk of its forces into three divisions, all of which were deployed forward, very near to the Israeli–Egyptian border adjacent to northeastern Sinai and Gaza.[12] All three divisions were optimized for an offensive maneuver warfare campaign. Only a handful of independent brigades and battalions were designated as a front reserve, to be parceled out to the divisions as necessary or to tackle separate assignments. The IDF had not prepared any frontier defenses to speak of.

The Egyptian army deployment on the eve of the Six-Day War constituted essentially a mirror image of the IDF deployment. Though it, too, had developed an operational plan for a preemptive strike against Israel as part of an offensive maneuver warfare campaign, the Egyptian government eventually decided to await an Israeli strike and then to counterattack after IDF forces had shattered themselves on the rocks of the Egyptian army's defensive positions.[13] To defeat the Israeli attack, the Egyptian army arrayed its forces in the Sinai in a traditional Soviet "sword and shield" layout. Infantry divisions, supported by armor and artillery, dug in behind stout defensive positions throughout northeastern Sinai and Gaza.[14] In the rear, to the west and south, the Egyptian army's armored and mechanized divisions readied themselves to counterattack Israeli forces once the latter had been worn down by Egyptian forward defenses.

On the morning of June 5, the first day of the war, IDF forces on the Sinai front numbered about 70,000 troops and 800 tanks. Egyptian army forces numbered about 100,000 troops and 900 tanks.[15] The

Egyptian army also possessed even greater advantages in other types of equipment, such as APCs and guns. Though outnumbered in fighting men and fighting machines—and despite the standard rule of thumb, which contends that an attacker most outnumber a defender by at least 3:1 in order to achieve victory—the IDF was nevertheless confident that it would defeat its adversary.

On June 4, the Israeli government authorized the IDF to initiate hostilities against Egypt on the following day. Early on June 5, the IAF commenced Operation Focus, its objective to smash Egyptian airpower on the ground through carefully orchestrated strikes against air bases.[16] Despite the fact that Egypt fully expected Israel to launch a preemptive strike, Focus caught the Egyptian air force completely by surprise, because the IAF struck from an unexpected direction (west to east) at an unexpected hour (well after dawn). IAF fighter-bombers repeatedly struck 18 Egyptian air force bases, as well as radar stations and surface-to-air missile (SAM) batteries, inflicting a crushing blow to the Egyptian air force. The IAF destroyed approximately 300 aircraft, including most of the Egyptian air force's frontline fighter-bomber fleet and all of its long-range bomber fleet. IAF fighter-bombers also knocked 13 air bases completely out of action, along with numerous radar stations and SAM batteries.[17] The Egyptian air force, in short, had been virtually annihilated in the space of a few hours. After Jordan and Syria opened hostilities against Israel in response to the Israeli attack against Egypt, the IAF expanded Focus to include their air forces as well, inflicting similarly destructive blows on them. Total Arab aircraft losses during Focus amounted to approximately 400 aircraft, while IAF losses to Arab air defenses amounted to 18 fighter-bombers.[18] By nightfall on June 5, the IAF had achieved air superiority over all of its adversaries.

Shortly after Operation Focus got underway, the IDF commenced Operation Red Sheet, the latter's objective to smash the Egyptian army in northeastern Sinai and Gaza. The IDF fought three break-in battles during the first day and night of Red Sheet. In the northernmost battle, the 84th (IS) Armored Division—an all-arms division composed of tanks, mechanized and parachute infantry, artillery, and engineers—quickly sliced through Egyptian defenses in its sector, ripping apart the 7th and 20th (EG) Infantry Divisions, as well as their supporting armored and artillery forces, at minimal cost to itself.[19] In the center-most battle, the 31st (IS) Armored Division—a division composed almost exclusively of tanks—raced over what the Egyptian army considered to be impassable terrain. Elements of this division subsequently surprised elements of the 4th (EG) Armored Division, which had begun to move northeastward to support the 7th and 20th (EG) Infantry Divisions, leading to a tank-versus-tank engagement that the IDF won rather handily, thereby helping to secure the left flank of the 84th (IS) Armored Division and the right flank of the 38th (IS) Armored Division, which would wage the southernmost break-in battle at Abu Ageila / Um Katef.[20]

## The Second Battle of Abu Ageila / Um Katef

The 38th (IS) Armored Division, an all-arms division, consisted of the 14th (IS) Armored Brigade (plus the 226th (IS) Independent Armored Battalion), the 80th (IS) Parachute Infantry Brigade, and the 99th (IS) Infantry Brigade, as well as attached artillery, engineering, and reconnaissance forces.[21] The 2nd (EG) Infantry Division—which included the 10th (EG) Infantry Brigade, the 12th (EG) Infantry Brigade, and an independent infantry battalion, as well as attached armor and artillery forces—garrisoned Abu Ageila / Um Katef.[22] The 12th (EG) Infantry Brigade, along with attached tank/tank destroyer and artillery forces, occupied the defensive complex itself, while the 10th (EG) Infantry Brigade, also with attached tank/tank destroyer and artillery forces, occupied a second defensive complex at Quseima, a few miles to the southeast of the Abu Ageila / Um Katef complex, but close enough so that it could participate in the defense of its sister complex if necessary.[23]

Shortly after the commencement of Focus, the lead elements of the 38th (IS) Armored Division crossed into the Sinai and headed toward Abu Ageila / Um Katef.[24] The battle that followed may be divided into two distinct phases: a preliminary phase, which occurred during the daylight hours of June 5, and a decision phase, which occurred during the nighttime hours of June 5–6. During the preliminary phase, the 38th (IS) Armored Division conducted several maneuvers. A mixed reconnaissance task force made up of armor, infantry, and artillery moved southward into the Sinai, eventually taking up positions between the Abu Ageila / Um Katef complex and the Quseima complex, albeit somewhat behind the pre-arranged schedule as a result of the difficult terrain in the area, its objective to keep an eye on the 10th (EG) Infantry Brigade in order to block any attempt by elements of this brigade to come to the assistance of the Abu Ageila / Um Katef complex. In the event, the 10th (EG) Infantry Brigade did not budge from its positions during the battle. Rather, it simply abandoned its positions after the fall of Abu Ageila / Um Katef and dissolved into a disorganized mass of men who fled into the interior of the Sinai, leaving behind their equipment.[25]

Meanwhile, under the cover of artillery fire and sporadic, small-scale air strikes (carried out primarily by armed training aircraft not suitable to participate in Focus), the 226th (IS) Independent Armored Battalion advanced northwestward into the Sinai in order to block Egyptian armored and mechanized reinforcements from arriving at Abu Ageila/Um Katef from northern or central Sinai, as well as to put itself into jumping-off position to launch an armored thrust against the rear of the complex.[26] Despite encountering stiff resistance along its route, as well as incurring some losses in fighting men and fighting machines, this armored battalion was in position to execute its assignment during the decision phase of the battle. While the 226th (IS) Independent Armored Battalion fought its way into position, so did the 14th (IS) Armored Brigade.[27] Also under the cover of artillery fire and sporadic, small-scale air strikes, it advanced more or less straight ahead from the Israeli border to the Abu Ageila/Um Katef complex, followed closely by engineering forces, which cleared the path forward of mines so that the 99th (IS) Infantry Brigade and the 38th (IS) Armored Division's artillery forces could move into their positions for the decision phase of the battle. Here, too, Israeli forces encountered stiff resistance at a forward strongpoint, but they were able to overcome it. By nightfall, the 14th (IS) Armored Brigade, the 99th (IS) Infantry Brigade, and the 38th (IS) Armored Division's artillery forces were in their designated jumping-off positions, ready to begin the decision phase of the battle.

This phase of the battle began shortly after sunset, when IAF transport helicopters inserted a much understrength battalion from the 80th (IS) Parachute Infantry Brigade into a "dead zone" of sand dunes inside the Abu Ageila/Um Katef complex.[28] The parachute infantry battalion then made its way during the next few hours over the sand dunes to its jumping-off position in the rear of the Egyptian artillery forces, its objective to knock out the artillery batteries. Under the cover of a heavy artillery barrage on the complex—the largest artillery barrage in IDF history up to that point in time—the parachute infantry battalion surprised the Egyptian artillerymen in their positions. In the close-range firefights that ensued, the Israeli parachute troops overcame their adversaries and largely silenced the Egyptian artillery. The parachute troops also thoroughly shot up a supply convoy inside the complex, adding further to the Egyptian garrison's confusion and panic.

Meanwhile, under the cover of the same artillery barrage, the 99th (IS) Infantry Brigade launched an assault against a series of trenches guarding the approach to the Abu Ageila/Um Katef complex.[29] While the Egyptian garrison was aware of the presence of the 226th (IS) Independent Armored Battalion and the 14th (IS) Armored Brigade in its vicinity as a consequence of the afternoon skirmishes, it was unaware of the presence of the 99th (IS) Infantry Brigade. Taking Egyptian infantrymen by surprise, Israeli infantrymen quickly broke into the trenches at their northern ends and began to flush them out moving southwards. Despite facing stiff resistance at points, the 99th (IS) Infantry Brigade succeeded in occupying the trenches without too much difficulty. Engineering forces followed hard on the heels of the 99th (IS) Infantry Brigade, clearing mines and antitank obstacles in front of the trenches, thereby permitting elements of the 14th (IS) Armored Brigade to come up in support of the 99th (IS) Infantry Brigade assault. Within a few hours, 99th (IS) Infantry Brigade and 14th (IS) Armored Brigade forces had penetrated into the heart of the Abu Ageila/Um Katef complex, sweeping away the Egyptian defenders before them.

The "final act" of the "drama" involved an attack by the 226th (IS) Independent Armored Battalion.[30] In its initial penetration of the complex, this armored battalion first overran 12th (EG) Infantry Brigade headquarters (HQ), adding yet another layer of confusion and panic to the Egyptian defense. Its primary objective, however, was to prevent Egyptian armored forces in the complex from coming to the assistance of Egyptian infantry forces. In tandem with elements of the 14th (IS) Armored Brigade moving to assist it, the 226th (IS) Independent Armored Battalion rather quickly got the upper hand against the Egyptian armored forces in an intense and close-range tank-versus-tank/tank destroyer engagement.

The Second Battle of Abu Ageila/Um Katef drew to an end by the morning of June 6. The Israeli victory had been complete; the Egyptian defeat total. Not only had the 38th (IS) Armored Division captured the entire Abu Ageila/Um Katef complex—the strongest defensive complex in all of the Sinai—but also it had done so at the limited cost of 30–40 men killed in action (KIA) and 20 tanks put out of action.[31] The battle had cost the Egyptian army an estimated 1,000 KIA and 40 plus tanks/tank destroyers put out of action.[32] Moreover, the 2nd (EG) Infantry Division had essentially ceased to exist, its soldiers retreating into the desert in an effort to reach the Suez Canal in order to cross back into Egypt proper. And all of the equipment possessed by the division that had not been destroyed during the battle fell into Israeli hands.

## Analysis of the Second Battle of Abu Ageila / Um Katef

An analysis of the battle in light of the six variables follows.

### *Operational Planning*

The IDF went into the battle very well prepared from the perspective of operational planning. First, from 1956–67, IDF officers routinely participated in war games whose objective was to capture the Abu Ageila / Um Katef defensive complex.[33] These war games concentrated less on identifying and rectifying the problems that the IDF had encountered in the First Battle of Abu Ageila / Um Katef during the Sinai Campaign than on the most efficacious means for overcoming this complex in its current configuration. Senior and junior IDF officers who participated in the second battle, therefore, were intimately acquainted beforehand with the IDF's strengths and weaknesses vis-à-vis those of the Egyptian army at Abu Ageila / Um Katef.

Second, based on an understanding of these relative strengths and weaknesses, the 38th (IS) Armored Division commander devised a meticulous and realistic operational plan based on the principles of traditional CAW.[34] Armor, parachute and nonmechanized infantry, artillery, and engineers would all exploit their particular strengths in order to compensate for the weaknesses of their sister arms during orchestrated attacks against specific vulnerable points of the Abu Ageila / Um Katef complex. The division commander also opted to execute the decision phase of the battle at night, because night fighting was at that time a strength of the IDF and a weakness of the Egyptian army. Furthermore, the 38th (IS) Armored Division commander devised a clever deception plan involving the actions of independent IDF forces stationed to the division's south, which had the effect of drawing the Egyptian army's attention away from the actions of his division.

The Egyptian army, too, was not idle from 1956–67 in respect of the Abu Ageila / Um Katef complex.[35] Based on the lessons that it drew from the First Battle of Abu Ageila / Um Katef, in tandem with its acceptance of Soviet war-fighting doctrine, the Egyptian army considerably expanded, modernized, and reinforced its defenses in and around the complex. In contrast to the IDF, however, the Egyptian army appears not to have made a determined effort to anticipate—let alone plan for—its adversary's tactics during the next round of hostilities. Hence, the insertion of a battalion of the 80th (IS) Parachute Infantry Brigade inside the complex, as well as the attack of the 99th

(IS) Infantry Brigade against the complex's trench system, took the Egyptian army wholly by surprise—with fatal results.

### *Intelligence and Reconnaissance*

Perhaps in part because of its sobering experience in the First Battle of Abu Ageila/Um Katef and perhaps in part because it recognized that it would almost certainly fight another battle here during a future war, the IDF thoroughly surveyed, mapped, and photographed the entire complex and its environs before withdrawing from the Sinai in the aftermath of the Sinai Campaign.[36] Furthermore, once the Egyptian army adopted the Soviet war-fighting doctrine, the IDF thoroughly familiarized itself with Soviet doctrine, including the Soviet approach to constructing and fortifying defensive positions.[37] IDF military intelligence also regularly kept tabs on the status of the complex over the years, collecting information through human intelligence (HUMINT), signals intelligence (SIGINT), and photographic intelligence (PHOTINT). While the IDF did not possess an exact picture of the Abu Ageila/Um Katef defensive complex on the eve of the Six-Day War—for example, it did not know that the Egyptian army had created only two lines of trenches rather than the standard three called for by Soviet war-fighting doctrine—it did possess a generally accurate picture of what it would be up against in respect of the Egyptian army's order of battle (ORBAT) and fortifications there.

Once the hostilities got underway, IDF forces engaged in continuous scouting activities. A reconnaissance task force from the 38th (IS) Armored Division kept the Egyptian garrison at the Quseima complex under constant surveillance. The IAF dispatched reconnaissance helicopters to help the 226th (IS) Independent Armored Battalion to locate and navigate over a route that enabled it to outflank the Abu Ageila/Um Katef complex, as well as to assist the 80th (IS) Parachute Infantry Brigade to identify the landing zone within the complex.[38] Other reconnaissance forces assisted the 14th (IS) Armored Brigade and the 99th (IS) Infantry Brigade to carry out their assignments during the battle.

The Egyptian army intelligence and reconnaissance effort constituted a mirror image of the IDF effort. Egyptian army intelligence apparently made no organized attempt to familiarize itself with the evolution of the IDF's force structure and war-fighting doctrine after the Sinai Campaign, either in general or as they pertained specifically to a possible future assault on the Abu Ageila/Um Katef complex. Undoubtedly, a paucity of HUMINT, SIGINT, and PHOTINT capabilities with which to monitor the IDF explains in large part this failure.

Once the hostilities got underway, the Egyptian army's only reconnaissance activities consisted of occasional reports about the movements of IDF forces sent in by observers in forward strongpoints guarding the approaches to the complex. After the IDF had reduced these strongpoints in the preliminary phase of the battle, Egyptian army forces in Abu Ageila/Um Katef were essentially left deaf and blind during the decision phase of the fighting. At no stage of the battle did the Egyptian army forces dispatch any mobile reconnaissance assets to keep tabs on IDF forces.

### *Exploitation of Terrain*

As in 1956, the natural terrain in and around the Abu Ageila/Um Katef complex strongly favored the defender.[39] As in 1956, the complex itself was built atop a plateau and "guarded" by hills along its southern flank and sand dunes along its northern flank, leaving no easy approach for an attacker from these directions. Moreover, several prominent hills to the east and northeast of Abu Ageila/Um Katef screened the complex from these directions, again leaving no easy approach for an attacker.

Already a quite formidable defensive complex during the Sinai Campaign, as the IDF discovered to its chagrin, Abu Ageila/Um Katef underwent a massive upgrade during the mid-1960s at the hands of the Egyptian army under the tutelage of Soviet advisors.[40] Two lines of continuous and fortified trenches, including antitank guns, were intended to impede an attack from the north. In front of these trenches lay a thick minefield, barbed wire, and antitank obstacles. Behind the trenches lay an artillery park—its batteries connected by a trench system—and, further back, an armor park, along with additional infantry fighting positions. The Egyptian army also fortified the hills to the east and northeast of the complex. On the eve of the Six-Day War, therefore, the Egyptian army appeared confident that it could hold the Abu Ageila/Um Katef complex against an IDF assault.

The IDF, however, devised clever means to overcome what appeared to be insurmountable terrain. A battalion of the 80th (IS) Parachute Infantry Brigade engaged in a vertical flanking maneuver, landing in and marching through supposedly impassable sand dunes in order to overrun the Egyptian army's artillery park. The 99th (IS) Infantry Brigade infiltrated silently through sand dunes, taking the 12th (EG) Infantry Brigade defenders by surprise in their trenches. And the 226th (IS) Independent Armored Battalion made its way through supposedly impassable sand dunes in order to attack Egyptian forces from the rear. The militarily sensible effort of the Egyptian army to upgrade the Abu Ageila/Um Katef defensive complex notwithstanding, the IDF managed to turn terrain disadvantages in the area into terrain advantages during the Second Battle of Abu Ageila/ Um Katef.

## *Quantity and Quality of Forces*

Even though the Egyptian army outnumbered the IDF in men and machines during the Six-Day War, the IDF did a far better job of concentrating its forces for battle, thereby giving it numerical superiority in certain engagements. The Second Battle of Abu Ageila/Um Katef constituted one such instance. The IDF assembled a total of approximately 19,000 men, 150 tanks, and 100 guns under the auspices of the 38th (IS) Armored Division.[41] The Egyptian army mustered a total of approximately 16,000 men, 90 tanks/tank destroyers, and 70 guns under the aegis of the 2nd (EG) Infantry Division. While the 38th (IS) Armored Division remained an essentially unified formation, the 2nd (EG) Infantry Division divided its manpower between the Abu Ageila/Um Katef and Quseima complexes, with approximately 8,000 men in each garrison. Thus, on the day of battle, the attacking IDF forces outnumbered the defending Egyptian army forces by an approximately 1.75:1 ratio in men (14,000 to 8,000), an approximately 1.5:1 ratio in armor (140 to 90), and an approximately 1.4:1 ratio in guns (100 to 70).

In respect of quality of forces, the equipment fielded by both the IDF and the Egyptian army at Abu Ageila/Um Katef gave each protagonist certain advantages vis-à-vis the adversary. The newer British, American, and French tanks in Israeli hands were better than the older models of Soviet tank/tank destroyer in Egyptian hands.[42] The guns, antitank weapons, and small arms in Egyptian hands, to the contrary, were better than the guns, antitank weapons, and small arms in Israeli hands. Overall, then, the quality of the equipment in the hands of each protagonist granted neither adversary a crucial advantage in the battle.

The relative quality of Israeli and Egyptian manpower, however, accorded the IDF a crucial advantage over its adversary.[43] Simply put, the rank-and-file Israeli soldier received better training than the rank-and-file Egyptian soldier, and he maintained and handled his arms better than his counterpart. In the weeks between the mobilization of armies and the outbreak of the Six-Day War, moreover, the 38th (IS) Armored Division conducted repeated training exercises associated with the capture of the Abu Ageila/Um Katef complex.[44] The 2nd (EG) Infantry Division, on the other hand, apparently did not conduct any training exercises related to the defense of the complex.

The IDF also had better leadership than the Egyptian army. From the lowest-ranking noncommissioned officer (NCO) all the way up the chain of command, IDF officers proved to be far less dependent on guidance from their superiors, far more aggressive in pursuit of their objectives, and far more flexible and innovative in achieving those objectives than their Egyptian counterparts.

## *Command and Control*

During the Six-Day War, the IDF's command and control ($C^2$) system was based upon a principle similar in nature to the German principle of Auftragstaktik (mission tactics). GHQ determined the basic objective of an operation, but then it deferred to commanders in the field with respect to how best to accomplish that objective Likewise, each level of command in the field down the chain of command acted in the same way with respect to the level of command below itself. This decentralized $C^2$ system—which promoted the traits of autonomy, aggressiveness, and flexibility in the officer corps—allowed IDF forces to react much more quickly and effectively to developments on the battlefield than their adversaries, which fit nicely with the emphasis on offensive maneuver warfare.

IDF GHQ determined the basic objective of Red Sheet—the collapse of the Egyptian army in the Sinai. GHQ then left it up to the 84th, 31st, and 38th (IS) Armored Division commanders to determine how best to accomplish this objective in their respective sectors. The 38th (IS) Armored Division commander devised an intricate set-piece operation—the decision phase of which would occur after nightfall, despite the reservations of IDF GHQ—to capture the Abu Ageila/Um Katef complex, assigning specific missions to the division's armored, infantry, parachute infantry, artillery, and engineering forces.[45] He then left it up to the commanders of these forces and their subordinates to work out the best ways to execute their particular missions.

No battle ever goes precisely according to plan. But the battle to capture Abu Ageila/Um Katef unfolded very largely as designed, despite a few hiccups here and there, in part because of the autonomy, aggressiveness, and flexibility displayed by IDF junior officers involved in the fighting.

The rigid $C^2$ system of the Egyptian army also helps to account for the Israeli victory at Abu Ageila/Um Katef. In contrast to the rather decentralized IDF system, the Egyptian army $C^2$ system was highly centralized, with junior officers neither encouraged nor expected to act autonomously, aggressively, or creatively.[46] Rather, they were expected to adhere strictly to orders from above. Such a rigid $C^2$ system might have been effective in the static defensive battle waged by the 2nd (EG) Infantry Division at Abu Ageila/Um Katef, but only if the division commander had been up to the task, which was not the case. New to the Sinai, away from his command post during the early hours of the battle, and out of close contact with his subordinates throughout the rest of the battle (because he had set up his command post at the Quseima complex), the division commander proved to be an ineffective leader.[47] In the absence of firm leadership from above, Egyptian army junior officers at Abu Ageila/Um Katef typically responded as if paralyzed throughout the battle. The Egyptian armor commander did not make any sort of move to support Egyptian infantry and artillery forces.[48] Nor did the officers of the 10th (EG) Infantry Brigade make any sort of move to support the 12th (EG) Infantry Brigade. Defeat could be the only result of such ineffective and paralyzed leadership.

### *Combined Arms Warfare*

Both the Israeli defense establishment and foreign military experts attributed the IDF's spectacular victory in the Six-Day War to the crushing effects of airpower and armor on the Arab armies, obscuring the fact that traditional CAW carried the day for the IDF during the break-in battles on the Sinai front (as well as on the Judean/Samarian and Golan fronts). A leading American authority on CAW remarked that:

> The set-piece attacks conducted by teams of Israeli paratroops, infantry, artillery, and tanks to break open the Egyptian border defenses were forgotten in the euphoria over [an] armored dash to the Suez Canal. . . .[49]

Nowhere was traditional CAW more in evidence in the Six-Day War than during the decision phase of the Second Battle of Abu Ageila/Um Katef. The 38th (IS) Armored Division exploited the inherent mobility of the 80th (IS) Parachute Infantry Brigade in order to suppress the 2nd (EG) Infantry Division's artillery fire via a vertical flanking maneuver, thereby assisting the 99th (IS) Infantry Brigade and the 14th (IS) Armored Brigade during their joint assault on the Abu Ageila/Um Katef complex. Engineering forces contributed to the assault as well by clearing minefields and antitank obstacles in front of the complex for the passage of armor. And the 226th (IS) Independent Armored Battalion covered the assault of the 99th (IS) Infantry Brigade and the 14th (IS) Armored Brigade by attacking the 2nd (EG) Infantry Division's armor before it could intervene in the fighting in support of the 12th (EG) Infantry Brigade.

While the 2nd (EG) Infantry Division possessed the arms to wage traditional CAW, it simply did not do so. Its artillery forces were put out of action early in the fight, so this arm did not contribute much to the battle. And its armored forces neglected to come to the aid of the 12th (EG) Infantry Brigade, instead waiting in place to be attacked by the 226th (IS) Independent Armored Battalion, leaving Egyptian infantrymen to face the combined fighting power of the 38th (IS) Armored Division's infantry, armored, and artillery forces.

## Ramifications of the Second Battle of Abu Ageila / Um Katef

The virtual annihilation of the Egyptian air force during Focus, in tandem with the crushing defeats suffered by Egyptian ground forces during the three break-in battles of Red Sheet, created an atmosphere of chaos and panic at Egyptian army GHQ, resulting in a hasty order for the Egyptian army in the Sinai to retreat over the Suez Canal to Egypt proper.[50] Predictably, what was intended to be an organized withdrawal soon devolved into a mad dash for safety. Thus began the post-Red Sheet phase of the IDF offensive in Sinai, a phase that had not been mapped out in advance of the war.[51] During this phase of the offensive, IDF GHQ decided to utilize an improvised combination of airpower and armor to destroy as much of the Egyptian army as possible before it retreated to safety across the canal. To accomplish this objective, armor from each of the IDF's three armored divisions raced westward to capture the Mitla, Giddi, and Bir Gifgafah Passes through which Egyptian army forces would have to withdraw if they were to reach the Suez Canal. Once the Egyptian army's routes of retreat had been cut off, IDF tanks and IAF fighter-bombers would then smash its bottled up forces in detail.

The 38th (IS) Armored Division's victory at Abu Ageila / Um Katef opened up the vast central region of the Sinai to the IDF. In cooperation with the 8th (IS) Armored Brigade, an independent brigade operating in the area, the 38th (IS) Armored Division routed Egyptian army forces in the central region of the peninsula from June 6–8, inflicting heavy losses in fighting men and fighting machines, as well as driving the remnants of those forces northwestward toward the passes, which were held by elements of the 84th and 31st (IS) Armored Divisions, to be destroyed there by a combination of airpower and armor.

By June 9, the IDF had captured all of the Sinai, putting an end to the Six-Day War on this front. The Egyptian army had suffered 10,000–15,000 soldiers KIA, and it had lost 80–85 percent of its pre-war equipment in the Sinai.[52] Many hundreds of its tanks, APCs, and guns had been destroyed, while many hundreds of others had been seized by the IDF, eventually to be incorporated into the Israeli arsenal. While IDF losses were certainly not inconsequential for an army of its size—approximately 400 soldiers KIA and up to 120 tanks knocked out of action (many of them later retrieved and repaired, however)—they paled in comparison to the losses suffered by the Egyptian army. Rarely in military history had such a lopsided result been obtained in war.

# 4

# Operation Blossom

The 1967 Six-Day War ended in a resounding Israeli victory on every front. The Israel Defense Forces (IDF) swiftly crushed the armed forces of Egypt, Jordan, and Syria, essentially annihilating their air forces and largely eviscerating their ground forces in the process. Moreover, in direct consequence of its triumphs on the southern, central, and northern battlefields, the IDF captured all of the Sinai Peninsula and Gaza from Egypt, all of Judea and Samaria from Jordan, and a substantial portion of the Golan from Syria, respectively, thereby obtaining for Israel defensible borders and a measure of strategic depth for the first time in the country's history.

Known as a military organization traditionally composed more of "doers" than "thinkers," the IDF did not conduct an in-depth and nuanced analysis of the reasons for its victory in the Six-Day War.[1] Even had it been an organization inclined to do so, the renewed hostilities that began shortly after the war along its frontiers (with Egypt, Jordan, and Syria), not to mention within them (in Judea, Samaria, and Gaza), almost certainly would have served to divert its attention away from any such inquiry. Instead, based on the undeniably excellent results obtained by the Israel Air Force (IAF) and the IDF Armored Corps during the fighting, the Israeli defense establishment rather offhandedly concluded that the Six-Day War had been won as a result of the combined effects of airpower and armor. "Consciously or otherwise," observed a prominent American military analyst, "[the IDF] came to rely largely on the tank–fighter-bomber team for its victories."[2]

Consequently, the Israeli defense establishment entered the 1969–70 War of Attrition with a very inflated sense of what airpower could accomplish on the battlefield. The flawed assumption that the IAF possessed the capability to determine the outcome of the next round of hostilities would lead the Israeli government to make a fateful decision about the use of airpower during the war—a decision that would ultimately cost the country dearly in respect of its national security.

## Prelude to Operation Blossom

The rapid and decisive outcome of the Six-Day War initially engendered a sense of optimism in the Israeli government that the country might be able to forge "land-for-peace" arrangements with certain of its Arab foes. On the one hand, the government never displayed much willingness to return Judea and Samaria to Jordan, because of the great importance of these territories to Israel for reasons of both national security and Jewish history.[3] On the other hand, the government displayed a real willingness to return the Sinai to Egypt and the Golan to Syria in exchange for peace treaties.[4] The Israeli government's position on land-for-peace vis-à-vis Egypt and Syria soon hardened, however, in the face of those countries' continued unwillingness in the wake of the Six-Day War to reconcile themselves to the existence of a Jewish state.

Any doubts about the collective Arab stance in respect of Israel would be cleared up just a few months after the war, at the Arab League summit in Khartoum, a conference that gave birth to the infamous "three noes": no recognition of Israel, no negotiations with Israel, no peace with Israel.[5] For its part, the Egyptian government heartily endorsed the Arab world's stance. In reference to that government's attitude about the return of the Sinai, a senior Israeli diplomat at the time crisply summarized its position as "that which has been taken by force must be recaptured by force."[6]

Indeed, even before its participation in the Arab League summit, Egypt unleashed its army against the IDF in an effort to oust Israel from the peninsula. Mere weeks after the Egyptian army's dazed remnants had beat a hasty retreat across the Suez Canal, Israeli forces adjacent to the Sinai bank of the waterway came under fire from Egyptian forces on the other side of the canal.[7] Renewed hostilities in the form of artillery barrages, special operations forces (SOF) raids, and air strikes, reasoned the Egyptian government, would not only inflict a steady toll of casualties and damage that the IDF could not long sustain, but also would inevitably draw the superpowers—the United States and the Soviet Union—into the Arab–Israeli conflict to the benefit of Egypt. This combination of military and diplomatic pressure, thought the Egyptian government, would be sufficient to coerce the Israeli government into ordering a unilateral IDF withdrawal from the Sinai, thereby "liberating" the territory without making any concessions whatsoever to Israel. Superpower pressure alone, after all, had been sufficient to coerce the Israeli government into ordering a unilateral IDF withdrawal from the peninsula in the

**1** Armor played a major role in the IDF's victories in the Sinai Campaign, the Six-Day War, and the Yom Kippur War.

**2** Nowadays the museum of the IDF Armored Corps, the police fortress at Latrun was heavily fortified and expertly defended by the Arab Legion, withstanding several attempts by Israeli forces to overrun it.

**3** Israeli forces captured this Egyptian army tank destroyer at the First Battle of Abu Ageila/Um Katef.

**4** Mechanized infantry forces played an important role in the decisive IDF victory at the Second Battle of Abu Ageila/Um Katef.

**5** The decision to rely solely on airpower contributed to the defeat suffered by the IDF in Operation Blossom.

**6** The Battle of October 14 ended in a decisive IDF victory, in part because of the close cooperation among armor, mechanized infantry, and artillery.

**7** Operation Mole Cricket 19 also ended in a decisive victory for the IDF, in part because the IAF made superb use of its UAVs to gather intelligence before and during the operation.

**8** IDF parachute infantry forces played a key role in blocking PLO forces from escaping the battlefield during Operation Hell.

**9** The actions of Shayetet 13 members contributed significantly to the success of Operation Spring of Youth.

wake of the 1956 Sinai Campaign. The Egyptian government thought that history would surely repeat itself.

From the summer of 1967 to the spring of 1969, hostilities between the Egyptian army and the IDF followed a rather routine pattern. The former would strike targets of opportunity along the eastern bank of the Suez Canal, intending to inflict serious damage on Israeli forces. The latter would respond to these provocations by hitting targets of opportunity on the western bank of the Suez Canal, intending to restore calm to the front. The IDF would also occasionally strike much deeper into Egypt, when the Israeli government deemed it necessary to do so in order to stabilize the front. In late 1968, for instance, IDF SOF destroyed economic infrastructure targets in the Nile Valley, including two bridges and a power station.[8] The IDF would also occasionally threaten much more valuable economic infrastructure targets, such as the Aswan High Dam, by activating the IAF to conduct reconnaissance flights over them.[9] The blows delivered by the IDF were adequate to calm the front for a while, but their deterrent effect would eventually fade, leading to yet another round of tit-for-tat skirmishes.

By the spring of 1969, the Egyptian government had arrived at two basic conclusions. First, Egyptian efforts to inflict serious harm on IDF forces and to compel superpower intervention in the region up to that point in time had not borne the desired fruit. Second, Egypt must therefore escalate the fighting from intermittent, low-level hostilities to sustained, high-intensity hostilities in order to accomplish these objectives and, hence, liberate the Sinai. Consequently, in March of 1969, the Egyptian government formally repudiated the cease-fire agreement that had ended the Six-Day War and announced the beginning of the War of Attrition.

This war would progress through four separate stages.[10] During the first stage of hostilities, which lasted from March 9–July 19, the fighting largely mimicked that which had occurred over the previous two years. The Egyptian army engaged in artillery barrages and SOF raids to which the IDF responded with artillery barrages and SOF raids of its own. This stage of hostilities, however, differed markedly from the fighting that had preceded it in respect of duration and intensity. IDF counterblows no longer had any deterrent effect on the Egyptian army's actions, leading to continuous and heavy exchanges of fire between the adversaries.[11]

Predictably, IDF casualties mounted steadily under the unrelenting Egyptian fire. Even the completion of the so-called Bar Lev Line—a string of fortified observation posts along the length of the Suez Canal, which were intended in part to shield soldiers on the front line from artillery and small arms fire—could not stem the rise in Israeli casualties.[12] To counter effectively the Egyptian army's fire, determined the Israeli government, it would be necessary for Israel to "escalate [the war] for the sake of de-escalation" in the words of the IDF Chief of Staff.[13] The government assumed that a massive IAF campaign against Egyptian army positions along the canal not only would cut down on IDF casualties, but also would inflict enough pain on Egyptian forces to convince the Egyptian government to halt the War of Attrition.

For the first time since the end of the Six-Day War, therefore, the Israeli government authorized the IAF to enter hostilities without restriction, signaling the start of the second stage of the war, which would last from July 20–December 31. Until July 20, the air force's contribution to the Israeli war effort had been largely confined to countering Egyptian air force incursions into Sinai airspace, conducting reconnaissance sorties inside Egyptian airspace, and transporting SOF into Egyptian territory. The limited participation of the IAF in the hostilities is indicated by the fact that, during the first stage of the war, its fighter-bombers shot down only 10 or so Egyptian fighter-bombers.[14]

From July 20–December 31, the IAF conducted a very intensive bombing campaign against Egyptian army positions along the Suez Canal, acting as "flying artillery" in order to make up for the IDF ground forces' glaring inferiority in artillery assets in relation to the artillery assets possessed by its Egyptian adversary. The Israeli air campaign aimed at the elimination of the Egyptian army's integrated air defense system (IADS), which consisted of a mix of surface-to-air missile (SAM) and antiaircraft artillery (AAA) batteries backed up by Egyptian air force fighter-bombers. The suppression of the Egyptian army's IADS would open the way for the suppression of the Egyptian army's artillery array, which had inflicted most of the IDF's casualties in the fighting.

By December 31, the IAF had essentially destroyed its adversary's IADS, had knocked out many of its adversary's artillery batteries, and had inflicted heavy casualties on its adversary's forces.[15] Its fighter-bombers had also destroyed another 35 or so Egyptian fighter-bombers in air-to-air engagements.[16] Nevertheless, despite its undisputed mastery of the airspace over the Suez Canal, the IAF could not quash Egyptian artillery fire, as the Soviet Union furnished additional artillery assets to Egypt at a faster pace than the IAF could destroy them. IDF casualties, as a result, continued to grow at an unacceptable rate for a country with such a small population as Israel. Moreover, the Egyptian government gave no sign that Egyptian army losses would compel Egypt to put an end to the war.

## Operation Blossom

This gloomy reality convinced the Israeli government that it needed to escalate the hostilities yet again in order to place even more pressure on Egypt to bring a halt to the fighting. From January 1–April 18, 1970, the War of Attrition progressed through its third stage. Not only would the IAF continue to attack Egyptian army positions along the Suez Canal during this stage, but also it would initiate a "deep-penetration" bombing campaign, known as Operation Blossom, against military targets in the Egyptian hinterland.[17] The Israeli government reasoned as follows in authorizing Blossom: deep-penetration air raids against targets in close proximity to major cities would bring home the grim reality of the war to the Egyptian populace, which would consequently lose faith in the Egyptian government, leading to the regime's weakening. Indeed, the Egyptian government might even fall and be replaced by a more moderate regime. In either case, Egypt would be compelled to bring the War of Attrition to a close on terms acceptable to Israel.[18] Moreover, the Israeli government believed that the United States would tacitly support Blossom, while the Soviet Union would not openly contest the operation by heavily involving itself directly in the hostilities.[19]

The IAF carried out 34 separate air raids as part of Blossom, involving a total of 118 sorties.[20] The accompanying table (see page 63) lists the particulars of the raids, all of which were executed by small formations of fighter-bombers, usually working in pairs or quartets. All of the targets were military in nature—army bases (including logistical and headquarters (HQ) facilities), SAM batteries, and radar stations. Almost all of the raids were carried out by just two understrength squadrons, the 69th and the 201st (IS) Squadrons, the only two IAF squadrons then equipped with state-of-the-art, American-supplied fighter-bombers with the capability to haul a large load of ordnance over long distances.[21] Indeed, the two squadrons between them had no more than 16 aircraft at the outset of Blossom.[22] Approximately two-thirds of the raids were carried out in January and February. These earlier raids were distinguished on the whole by their close proximity to major cities. A couple of examples suffice to provide a general portrait of the first batch of raids. On January 13, IAF fighter-bombers struck an army base at Khanka, 12.5 miles (20 km) from Cairo; on

January 28, they struck an army base at Ma'adi, a mere 6 miles (10 km) from Cairo.[23] The later raids in March and April concentrated primarily on SAM batteries and radar stations in the Nile Delta, as opposed to army bases adjacent to cities.[24] Again, a couple of examples suffice to provide a general portrait of the second batch of raids. On March 23, IAF fighter-bombers struck a radar station in the Nile Delta; on March 31, they struck a SAM battery in the same region.[25]

Tactically speaking, the deep-penetration raids constituted a clear win for Israel, as they accomplished their objectives. IAF fighter-bombers arrived at their targets unchallenged by Egyptian air defenses, damaged or destroyed those targets, and returned to Israeli airspace without loss.[26] Moreover, the raids sometimes inflicted very heavy casualties. The two that comprised Blossom 5, for instance, apparently caused the deaths of over 100 Egyptian army personnel and nine Soviet army personnel, at least two of whom were senior officers acting in an advisory capacity.[27] On those rare occasions when Egyptian aircraft were able to intercept IAF fighter-bombers during their egress from the targets, the Egyptians came up the losers.[28] All in all, the Egyptian government and populace experienced a profound sense of humiliation with respect to the ease with which IAF fighter-bombers flew far into Egyptian airspace to attack targets in the country's hinterland.

Strategically speaking, to the contrary, the deep-penetration raids constituted an utter fiasco for Israel, as they boomeranged on the country. The actual results of the deep-penetration raids at this level turned out to be precisely the opposite of the predicted results.[29] First, despite its sense of humiliation, the Egyptian populace did not turn on the Egyptian government. Rather, it rallied around the regime. Second, the Egyptian government displayed no signs that it would end the War of Attrition in response to the deep-penetration raids. The fighting along the Suez Canal remained as heavy and as sustained during January–April 1970 as it had been during July–December 1969, and the IDF continued to absorb casualties at an alarming rate. Third, the United States declined to back the deep-penetration raids, most emphatically by its refusal to furnish the numbers of aircraft requested by Israel.[30] And, fourth, the Soviet Union seized on the raids as a pretext to insert its forces openly and extensively into the War of Attrition.[31] This confluence of disadvantageous developments placed Israel in a very difficult spot as the country entered the last stage of the war.

## IAF Deep-penetration Raids*

| *Blossom Number* | *Raid Date* | *Designated Squadron(s)* | *Assigned Target(s)* |
|---|---|---|---|
| 1 | January 7 | 69 | Army Base |
| 1 | January 7 | 201 | Army Base |
| 2 | January 13 | 115 | Army Base |
| 2 | January 13 | 115 | Army Base |
| 2 | January 13 | 109 | Army Base |
| 2 | January 13 | 201 | Army Base |
| 3 | January 18 | 69 | Army Base |
| 3 | January 18 | 201 | Army Base |
| 4 | January 23 | 201 | Army Base |
| 5 | January 28 | 69 and 201 | Army Base |
| 5 | January 28 | 69 | Army Base |
| 6 | February 2 | 69 | Radar Station |
| 6 | February 2 | 201 | Army Base |
| 7 | February 6 | 201 | Army Base |
| 7 | February 6 | 69 | Army Base |
| 8 | February 8 | 201 | Army Base |
| 8 | February 8 | 69 | Army Base |
| 9 | February 12 | 201 | Radar Station |
| 9 | February 12 | 69 | Army Base |
| 9 | February 12 | 201 | Army Base |
| 9 | February 12 | 69 | Radar Station |
| 10 | February 17 | 201 | SAM Battery |
| 11 | February 26 | 201 | SAM Battery |
| 11 | February 26 | 69 | SAM Battery |
| 12 | March 6 | 69 and 201 | Radar Station/ SAM Battery |
| 13 | March 13 | 201 | Radar Station |
| 14 | March 17/18 | 69 and 201 | Raids Aborted |
| 15 | March 23 | 69 | Radar Station |
| 16 | March 26 | 69 | SAM Battery |
| 17 | March 31 | 69 and 201 | SAM Battery |
| 18 | April 3 | 69 | SAM Battery |
| 19 | April 8 | 201 | Army Base |
| 20 | April 10 | 201 | Radar Station |
| 21 | April 13 | 201 | Radar Station |
| 21 | April 13 | 69 | SAM Battery |

* This table is based on data compiled from Yaacov Bar-Siman-Tov, *The Israeli–Egyptian War of Attrition, 1969–1970: A Case Study of Limited Local War* (New York: Columbia University Press, 1980), pp. 134–135; the official IAF website, <www.iaf.org.il>; and Danny Shalom, *Phantoms Over Cairo: The Israeli Air Force in the War of Attrition (1967–1970)*, Vol. B (Rishon LeZion: Bavir Aviation & Space Publications, 2007). In instances of disagreement among the sources over dates and targets, the author has included the most reliable data in the table.

## Analysis of Operation Blossom

An analysis of the operation in light of the six variables follows.

### *Operational Planning*

From a tactical standpoint, the IAF engaged in meticulous preparations for the deep-penetration raids. The air force chose the targets of its raids with considerable care not only in an effort to maximize the psychological effect of in-depth bombing on the populace and the regime, but also in an attempt to minimize the amount of collateral damage inflicted on civilians and nonmilitary infrastructure. The IAF plotted flight routes to and from the targets based on accurate and up-to-date information about the dispositions and capabilities of Egyptian air defense assets, which is an important reason why most of the raids went unopposed by the Egyptian army. The professionalism shown by IAF planners helps to account for the success of the deep-penetration air raids at the tactical level.

From an operational standpoint, IDF and Israeli government staff work displayed both hastiness and sloppiness of thought, emerging as a virtual mirror image of IAF staff work. First, neither the IDF nor the Israeli government apparently gave any consideration whatsoever either to the past or the contemporary record of bombing when employed as a psychological instrument to influence an adversary's conduct. If they had done so, they would undoubtedly have discovered that the massive, years-long Allied bombing campaigns against Germany and Japan during the Second World War—the tangible effects of which on industry and other infrastructure certainly did assist Allied forces in their eventual victories over their Axis adversaries—did not undermine the resolve of either the latter's populaces or regimes to continue the war.[32] If further evidence were necessary to reveal that bombing was not an efficient means by which to undermine an adversary's will, the IDF and Israeli government could also have considered the Vietnam War, a conflict in which widespread and heavy American bombing did not undermine the resolve of the North Vietnamese populace or regime to carry on the fight. This lack of attention to past and contemporary history encouraged the IDF and Israeli government in their rather silly belief that a relative handful of small-scale air raids would be sufficient to coerce Egypt into putting an end to the War of Attrition.

Second, but certainly no less important, while planning Blossom, neither the IDF nor the Israeli government apparently considered what might happen should either the United States or the Soviet Union react to the operation in an unanticipated manner. When both the United States and the Soviet Union responded to Blossom with unexpectedly hostile moves, both the IDF and the Israeli government were at a serious loss as to how to deal with this new and very unwelcome situation.

For their part, from a tactical standpoint, neither the Egyptian army nor the Egyptian government engaged in any effective preparations in order to thwart Israeli deep-penetration air raids; however, from an operational standpoint, the Egyptian government showed itself to be a shrewd manipulator during the third stage of the War of Attrition in respect of soliticing outside intervention in the hostilities, reinforcing cleverly in Soviet minds the Soviet government's existing inclination to intervene in the fighting.[33]

### *Intelligence and Reconnaissance*

A distinction between the tactical and operational levels also needs to be drawn in respect of this variable. From a tactical standpoint, except on rare occasions, Israel possessed accurate and up-to-date information throughout the War of Attrition not only about Egyptian army dispositions and capabilities, but also about Soviet military dispositions and capabilities in Egypt. A constant stream of data was furnished by a combination of IDF military intelligence Unit 848 (nowadays known as Unit 8200), which collected signals intelligence (SIGINT), IAF aerial reconnaissance sorties, which collected photographic intelligence (PHOTINT), as well as SOF forays (and civilian agents stationed in Egypt), which collected human intelligence (HUMINT).[34] The IAF could not have carried out the deep-penetration raids with such success in the absence of solid and current "battlefield" intelligence.

From an operational standpoint, on the other hand, Israeli military intelligence efforts proved to be inept. The intelligence community failed to assess correctly the Egyptian, Soviet, and American responses to IAF deep-penetration air raids. In the case of the Soviet Union, the intelligence community simply lacked a cadre of experienced "Sovietologists" who were able to evaluate insightfully Soviet behavior during the War of Attrition.[35] In the case of the United States, the intelligence community, like the government, appeared to rely far too heavily on a single source of information, the Israeli ambassador in Washington, in putting together its faulty judgment that America would stand behind Israel vis-à-vis the deep-penetration raids.[36]

For their part, neither the Egyptian army nor the Egyptian government predicted that Israel would embark on a deep-penetration bombing campaign in order to try to bring about the end of the War of Attrition. Nor did the Egyptian army collect any sort of tactical-level data in connection with Blossom.

## *Exploitation of Terrain*

From a tactical standpoint, IAF fighter-bombers took good advantage of the terrain to shield themselves from Egyptian air defenses during the deep-penetration raids, seemingly popping up from nowhere over their targets, by flying at very low altitudes, where Egyptian radar stations could not effectively track them, because of "ground clutter."[37] By way of contrast, no evidence has come to light to suggest that Egyptian forces attempted to make use of the terrain to their advantage. From an operational standpoint, however, Israeli superiority at the tactical level made no difference to the ultimate outcome of Blossom.

## *Quantity and Quality of Forces*

From a tactical standpoint, the superior quality of Israeli forces vis-à-vis their Egyptian counterparts was plainly manifest in the ease with which IAF fighter-bombers carried out the deep-penetration raids. Egyptian air defenses could not prevent some of the best-trained and most experienced airmen in the world, who were flying among the most advanced fighter-bombers in the world, from executing their missions.

From an operational standpoint, however, the qualitative advantage possessed by Israel was nullified by its self-imposed quantitative disadvantage. Deploying a small number of fighter-bombers to conduct a small number of small-scale air raids over a four-month period was simply never going to be sufficient to coerce Egypt into calling a halt to the War of Attrition on terms acceptable to Israel.

## *Command and Control*

From a tactical standpoint, $C^2$ during the deep-penetration air raids proved to be effective. Communications relay aircraft were deployed in order to maintain contact between IAF HQ and the fighter-bombers of the 69th and 201st (IS) Squadrons at all times.[38] That the air raids unfolded essentially according to plan, notwithstanding a couple of mishaps, serves as a testament to IAF $C^2$ capabilities during the War of Attrition. The Egyptian army's $C^2$ system, on the other hand, could not effectively cope with the IAF raids. From an operational standpoint, however, superior Israeli capabilities at the tactical level, too, made no difference to the ultimate outcome of Blossom.

### *Combined Arms Warfare*

The IDF did not practice traditional combined arms warfare (CAW) during Blossom. The IDF and Israeli government entrusted the task of bringing Egypt to its knees solely to the IAF. Airpower alone, the IDF and the Israeli government convinced themselves, would be sufficient to achieve this objective. The IAF, in turn, entrusted the task to just two understrength squadrons of fighter-bombers. In hindsight, however, only a very large-scale, combined arms offensive across the Suez Canal would have stood at least a chance of obtaining the desired objective. Indeed, a former IAF commander later observed that:

> Instead of seeing the Air Force [sic] as one of the components in the fighting and putting in land forces and armor on a significant scale in order to crush the Egyptian land front with all its artillery and missiles, they [i.e., the IDF and the Israeli government] entrusted the entire task to the Air Force [sic] alone and saw this as the sole solution, while the land forces continued to dig themselves in. What was needed was a land operation on a significant scale that would have [had] a wide and deep impact.[39]

Neither the IDF nor the Israeli government, though, evinced any willingness throughout the War of Attrition to incur the substantial costs in blood and treasure that would almost certainly have resulted from a large-scale, cross-canal offensive.

## Ramifications of Operation Blossom

Discouraged that the deep-penetration air raids were not having the desired effect on Egypt, blindsided by the American refusal to support the raids, and wishing to avoid a head-on collision with Soviet forces over the raids, the Israeli government suspended them in mid-April, clearing the way for the final stage of the War of Attrition, which lasted from April 19–August 8. To wrest the initiative away from Israel, the Soviet Union had already deployed 6,500–8,000 soldiers and technicians, including aircrews and SAM crews, to Egypt before the deep-penetration raids drew to a close, a number that would balloon to at least 12,000 by June.[40] Emboldened by the Israeli retreat, Egyptian and Soviet forces, working in close cooperation, spent these months methodically rolling forward a new and more sophisticated IADS toward the Suez Canal, sparking some of the most severe fighting during the War of Attrition. The intensity of the hostilities throughout this stage of the war is demonstrated by the fact that the IAF shot down some 45 Egyptian air force fighter-bombers, including at least five flown by Soviet pilots (in a major clash over the Gulf of Suez).[41] To cite one particular instance of the fierce fighting typical of the final stage of the War of Attrition, the commander of the 69th (IS) Squadron later recalled that, during a single mission on July 18, eight aircraft from his squadron, along with eight more from the 201st (IS) Squadron, killed 13 Soviet army air defense personnel—and, undoubtedly, a much larger number of Egyptian army air defense personnel—in an air strike that destroyed several SAM batteries (at a cost of two fighter-bombers lost to missiles).[42] Despite inflicting very heavy losses in men and matériel on both Egyptian and Soviet forces during the final stage of the war, the IAF was ultimately unable to prevent them from establishing a very potent IADS adjacent to the Suez Canal.[43] Furthermore, the IAF lost a significant number of fighter-bombers to the new air defenses.[44]

Under American pressure to halt the hostilities, and with little hope that the IAF would regain air superiority over the Suez Canal, the Israeli government consented to a cease-fire in early August. A senior Israeli diplomat framed the government's decision as follows:

> In late July, when we [the government] surveyed our situation in all its aspects, we felt that the risks of accepting the American cease-fire proposal were far less than the dangers of rejecting it. Rejection would mean the continuation of savage war with Egypt, the prospect of involvement in military conflict with the Soviet Union, and diminishing American fidelity to Israel.[45]

While the War of Attrition did not result in an Israeli defeat—rather, the war ended in a stalemate, with Israel still in firm control of the Sinai—Blossom did lead to a major setback for the country, the loss of air superiority in the vicinity of the Suez Canal. This loss would be made even worse by flagrant Egyptian and Soviet violations of the cease-fire agreement in the months following the war in order to reinforce and upgrade their forces' canal-side IADS.[46] This loss of air superiority, coupled with the country's overall lack of preparedness for the outbreak of the 1973 Yom Kippur War, would cost the IDF very dearly in the Sinai during the opening days of that war. Indeed, it would take almost three weeks of long and bloody fighting for Israeli forces to overcome their initial reverses on this front.

# 5

# The Battle of October 14

The Israel Defense Forces (IDF) confronted its most severe challenge since the 1947–49 War of Independence during the 1973 Yom Kippur War. Caught unprepared for the outbreak of war, its formidable reserve armored divisions not deployed for battle, the IDF had to counter simultaneous and massive Syrian and Egyptian offensives on the Golan and in the Sinai, respectively, with only thin screens of armor, infantry, and artillery to stand in the way of its adversaries. The Syrian army held a roughly 5:1 advantage in tanks (an 8:1 advantage counting its second-echelon tanks), a 10:1 advantage in infantry, and a 12–13:1 advantage in artillery along the so-called Purple Line, the cease-fire line that separated Israeli and Syrian forces on the Golan.[1] The Egyptian army held a roughly 14:1 advantage in tanks, an 80:1 advantage in infantry, and a 40:1 advantage in artillery along the banks of the Suez Canal, the cease-fire line that separated Israeli and Egyptian forces.[2] Moreover, as the result of a stern admonition from the United States, in the hours just prior to the commencement of hostilities, to refrain from authorizing a preemptive air strike, the Israeli government forbade the IDF from delivering the first blow of the war, a blow that might have at least tilted the odds a bit less against Israeli forces.[3]

In consequence of this "perfect storm" of circumstances, the IDF suffered nearly catastrophic reverses on both fronts during the early days of hostilities. On the Golan, while a brilliant mobile campaign waged by a lone Israeli armored brigade stymied every Syrian effort to advance in the northern sector of the plateau, the Syrian army broke through Israeli defenses in the southern sector, threatening to reach Israel proper. In the Sinai, two Egyptian army corps crossed the Suez Canal, established bridgeheads on its eastern bank, overran most of the Bar Lev Line (a string of fortified Israeli strongpoints along the waterline), and inflicted heavy casualties on Israeli armored forces stationed near the canal.[4]

Nevertheless, the IDF was ultimately able to contain the Syrian and

Egyptian offensives, stabilizing both fronts. In the north, Israeli forces launched a counterattack on the third day of the war (October 8), cleared the Golan of Syrian forces by the fifth day of the war (October 10), launched a counteroffensive into Syria on the sixth day of the war (October 11), and advanced to within artillery range of the Syrian capital by the ninth day of the war (October 14). Hostilities would continue for another 10 days on this front, but the Syrian army (and its Arab allies) would make no headway in repelling Israeli forces from the gates of Damascus.

In the south, to the contrary, the IDF recovered its balance far more slowly.[5] Here, too, Israeli forces launched a counterattack on the third day of the war; however, unlike its northern equivalent, the counterattack in the south culminated in a costly and humiliating defeat for Israeli forces, which lost the better part of two tank battalions, as well as some additional territory in the Sinai.[6] For the next five days of hostilities (October 9–13), the IDF was content mainly to hunker down in defensive positions in order to re-equip and reorganize its forces, as well as to consider its options moving forward. Whereas the IDF had essentially defeated the Syrian army by the ninth day of the Yom Kippur War, it had yet to register any meaningful victory against the Egyptian army, locked as Israeli forces were in a disadvantageous standstill with their adversary.

This stalemate, though, would come to a swift end on October 14. In the space of a few hours during that day, the Egyptian army suffered a crushing defeat at the hands of the IDF in the single largest armored battle since the Second World War. Indeed, the armored battle of October 14 almost certainly ranks as the second largest such battle in the annals of warfare, surpassed in numbers of tanks engaged in the fighting only by the titanic clash of the German and Soviet armies at Kursk in 1943.

## Prelude to the Battle of October 14

On the fourth day of the war (October 9), Israeli diplomats in the United States informed their American counterparts that the IDF had lost approximately 500 tanks and 50 aircraft during the first three days of hostilities, noting that approximately 400 of the former had been lost in the Sinai.[7] On the first day of the war (October 6) alone, the IDF lost approximately 150 tanks (as many as 175 if mechanical breakdowns and accidents are added to the total) of the approximately 285 tanks that it had on hand in the Sinai at the outbreak of the war.[8] Much of this armor was knocked out of action while participating in desperate "tank charges" undertaken without support from infantry, artillery, or airpower to relieve Bar Lev Line strongpoints under siege, many of the tanks succumbing to the missiles and rockets wielded by specially trained and equipped Egyptian antitank infantry. The holding battles on the second day of the war (October 7) by the remnants of the regular armored division in the Sinai, the hurried deployment of two reserve armored divisions to the canal zone (many of the tanks had to make the journey on their own tracks, as the IDF did not possess enough tank transporters to accommodate such a rapid mobilization), as well as the botched counterattack by these reserve divisions on the third day, cost Israeli forces approximately two hundred more tanks out of action.[9] Indeed, one of the reserve armored divisions lost approximately 50 tanks in the space of a few hours during the counterattack, with Egyptian antitank infantry once again accounting for much of this armor.[10] Concomitantly, Egyptian air defenses shot down 19 Israel Air Force (IAF) fighter-bombers—four on October 6, eight on October 7, and seven on October 8—during the early days of the war, as well as damaged an unknown number of other aircraft.[11] Many of these aircraft were hit while conducting strikes against easily repairable floating bridges over the Suez Canal, the Egyptian army simply replacing the damaged components with new ones as necessary.

In light of the utter collapse of the counterattack in the south, as well as the steady progress being made by the counterattack in the north, IDF General Headquarters (GHQ), in consultation with the Israeli government, decided that the best course of action for the IDF to pursue from the fourth day of the war would be to press ahead on the Golan, in an attempt to knock Syria out of the war, while holding its ground in the Sinai; therefore, beginning on October 9, Israeli forces adopted an essentially defensive posture in the south.[12] For the next five days, both Israeli and Egyptian forces were content simply to probe each other's forward positions, leading to limited clashes each day in which the former routinely prevailed, because of the superior long-range gunnery and maneuver warfare capabilities of IDF armor.

The IDF used the days from October 9–13 mainly to replenish and reorganize its forces. Thanks to highly efficient maintenance crews, many damaged tanks were retrieved, repaired, and returned to service, while additional armored forces streamed into the Sinai, increasing the tank order of battle (ORBAT) by approximately 50 percent.[13] Simultaneously, the IDF integrated substantial infantry and artillery assets into its armored forces, refined its tactics, and renewed its commitment to combined arms warfare (CAW). The reckless, unsupported tank charges of the first three days of the war, which led to such grievous losses, would not be repeated in the future.[14]

Meanwhile, the IDF and the Israeli government debated Israel's next move. IDF GHQ, IDF Southern Command Headquarters (HQ), and the Sinai division commanders debated whether it would be better for Israeli forces to attempt an early counteroffensive across the Suez Canal or better to wait for Egyptian forces to attempt to break out of their canal-side bridgeheads, smash them in a mobile armored battle, and then mount a counteroffensive across the canal. The debate, at this juncture of the war, did not concern the best means to inflict a decisive defeat on the Egyptian army, but rather the best means to convince Egypt to accept a cease-fire. Even proponents of an early counteroffensive conceded that such a move should aim to accomplish the latter objective.[15] By the eighth day of the war (October 13), for its part, the Israeli government had consented, albeit reluctantly, to accept an American-backed proposal for a cease-fire, perhaps thinking that Israeli territorial gains in the north might be leveraged to offset territorial losses in the south during postwar diplomatic negotiations.[16] The Egyptian government, however, unequivocally turned down the proposal.

Why the Egyptian government would reject a cease-fire at this point in the Yom Kippur War is not so very hard to fathom: the Egyptian politico–military leadership thought that the Egyptian army was winning the war, and therefore concluded that it would be possible to achieve further gains on the battlefield. The Egyptian leadership had grounds for optimism. Not only had the Egyptian army put two army corps—comprised of five infantry divisions, as well as accompanying armor, artillery, engineering and air defense forces—across a major water obstacle and set up defensible bridgeheads on the first day of hostilities, but also it had broken the Bar Lev Line—all at the cost of approximately 200 dead during that first day (as opposed to a prewar estimate of up to 10,000 dead).[17] And, on the third day of the war, its forces had defeated a large-scale IDF counterattack against those bridgeheads, inflicting substantial losses on Israeli armor.[18]

Still, even as the Egyptian army basked in its early successes, ominous warning signals appeared on the horizon. First, its bridgeheads in Sinai remained rather shallow—indeed, in some places, the bridgeheads did not even extend as far inland as the very modest distances called for in the Egyptian war plan. Second, despite taking a heavy toll of Israeli fighting men and fighting machines, Egyptian losses exceeded those of its adversary in most—possibly even all—categories, often by significant margins, over the first eight days of the war.

Whereas the IAF had lost 19 aircraft in the hostilities through October 8, to cite one example, the Egyptian air force had lost approximately 80 aircraft, including many troop-carrying helicopters transporting special operations forces (SOF) behind Israeli lines, mostly in air-to-air engagements.[19] And, third, whenever Egyptian forces attempting to advance beyond their bridgeheads from October 9–13 ran into determined Israeli resistance, they were repulsed, often at high cost to themselves.

Why, then, did the Egyptian army launch a major offensive toward the Sinai's mountain passes, the gateways to the interior of the peninsula, on the morning of October 14? One school of thought contends that the offensive essentially constituted an act of altruism, that it was undertaken solely in response to a Syrian plea for assistance, the argument being that the attack was intended to relieve pressure on the Syrian army by diverting the IDF's attention southward. This school further contends that the final Egyptian war plan never envisaged a major offensive beyond the bridgeheads, that the Egyptian government's intention in crossing the Suez Canal was not to retake the Sinai via hostilities, but rather to break the diplomatic stalemate over the peninsula by compelling the superpowers to intervene in the war in order to begin a political process for the territory's return to Egypt.[20] To support its argument, this school of thought cites the testimonies of high-ranking Egyptian officers who claim to have been firmly opposed to the offensive, the lack of careful planning and preparations for the offensive (as opposed to the meticulous planning and preparations for the crossing of the canal), and the thoroughly disorganized and torpid implementation of the offensive itself.

This school's argument certainly has merit, as no doubt exists that the Egyptian government authorized the offensive in response to a Syrian plea; however, if the only reason for the attack was to relieve pressure on its northern ally, then why did the government refuse to accept a cease-fire on October 13, a cease-fire that Israel had already endorsed? A cease-fire, after all, would have served not only to relieve the pressure on Syria, but also to promote the Egyptian agenda in the Sinai, as a cease-fire would have secured the territorial gains there necessary to break the diplomatic stalemate.

In addition to coming to the aid of Syria, it seems, fidelity to Soviet military doctrine, which heavily influenced Egyptian war planning, buoyed by an overly optimistic assessment of the Egyptian army's performance in the first eight days of the war, also spurred the Egyptian government to command Egyptian forces to engage in the offensive.[21] Soviet doctrine, which called for an "operational pause" after the crossing to consolidate the canal-side bridgeheads, also called for a follow-on attack to break out of these bridgeheads in order to advance inland, and iterations of the Egyptian army GHQ's prewar plans incorporated just such a major offensive to capture the Sinai passes. However belatedly, and in spite of the putative opposition of some high-ranking officers, therefore, the Egyptian army was adhering to a script that had been written before the Yom Kippur War.

## The Battle of October 14

For the first seven days of the war, the Egyptian army refrained from deploying its armored divisions across the Suez Canal. Hundreds of Egyptian tanks had crossed into the Sinai during that time, but these tanks belonged to armored and mechanized brigades attached to the infantry divisions that had established the bridgeheads across the canal. On October 12, the IDF received intelligence reports indicating that the Egyptian army had begun to make preparations to transfer its armored divisions to the Sinai. On October 13, the IDF received intelligence reports that elements of these divisions had actually begun to cross over the canal.[22]

Well aware of the nuances of Soviet military doctrine, IDF GHQ recognized that the operational pause phase of the Egyptian war plan was ending and that the breakout offensive phase was beginning. Consequently, with the consent of the Israeli government, IDF GHQ decided that Israeli forces would first smash the Egyptian offensive and only then initiate a counteroffensive across the canal.[23] Israeli forces in the Sinai, in turn, were instructed to get ready to meet the Egyptian offensive.

The Egyptian offensive called for five separate armored thrusts: two main penetration, two diversionary flanking, and one blocking.[24] The main armored thrusts, one from the northern bridgehead and one from the southern bridgehead, were to break through the Bir Gifgafah and Mitla Passes, respectively, and link up in a deep pincer movement behind Israeli lines, removing the IDF's hold on the high ground to the east of the Suez Canal and, potentially, opening up the interior of the Sinai to further Egyptian conquests. The northern diversionary thrust was to advance toward Romani, near the Mediterranean Sea coast, while the southern diversionary thrust was to advance toward Ras Sudar in southwestern Sinai, drawing Israeli attention and assets away from the main thrusts. Finally, a blocking thrust from the southern bridgehead was to ensure that Israeli forces in the vicinity of the Gidi

Pass could not attack the vulnerable flank of the main thrust advancing toward the Mitla Pass.

To accomplish its objective, the Egyptian army assembled an impressive ORBAT.[25] The 2nd (EG) Army Corps, responsible for the northern bridgehead, had three infantry divisions—the 18th (EG) Infantry Division, the 2nd (EG) Infantry Division, and the 16th (EG) Infantry Division (from north to south)—already in the Sinai, each with an attached armored brigade. Two additional brigades of the 21st (EG) Armored Division—one armored and one mechanized—were transferred to the northern bridgehead in anticipation of the offensive, while the division reincorporated its second armored brigade, which had been fighting in the Sinai under the auspices of the 16th (EG) Infantry Division. The 21st (EG) Armored Division, in other words, was now at full strength in the Sinai. The 3rd (EG) Army Corps, responsible for the southern bridgehead, had two infantry divisions—the 7th and 19th (EG) Infantry Divisions (from north to south)—already in the Sinai, each with an attached armored brigade. An additional armored brigade from the 4th (EG) Armored Division was transferred to the southern bridgehead in anticipation of the offensive. All told, the Egyptian army had amassed up to 1,000 tanks in the Sinai for its breakout offensive, parceled out among seven armored brigades and supporting infantry forces (though not all of these tanks would participate in the fighting on October 14).

To counter the Egyptian offensive, the IDF also fielded an impressive ORBAT.[26] In the north of the 2nd (EG) Army Corps sector, opposite the 18th (EG) Infantry Division, the IDF deployed "Force Tiger," which included two mechanized infantry brigades with the equivalent of two armored battalions. In the center of this sector, opposite the 2nd (EG) Infantry Division sector, the 162nd (IS) Armored Division deployed one armored brigade, which had been assigned to the division for the duration of the Egyptian offensive. Because the 162nd (IS) Armored Division had been designated by IDF GHQ to be the first Israeli force to cross over the Suez Canal during the planned Israeli counteroffensive, its three armored brigades had been pulled off of the front line to prepare for the crossing operation. They would also serve as a reserve force in case Egyptian forces threatened to break through the Israeli front line. In the south of this sector, opposite the 16th (EG) Infantry Division sector, the 143rd (IS) Armored Division deployed two armored brigades, with a third armored brigade in reserve. And, in the 3rd (EG) Army Corps sector, opposite the 7th and 19th (EG) Infantry Divisions, the 252nd (IS) Armored Division deployed two armored brigades, one mechanized infantry brigade, and elements of a parachute infantry battalion. All told, the IDF had amassed approximately 750 tanks in Sinai, approximately 450 of these tanks on the front line and 300 in reserve.

In the early morning of October 14, the Egyptian army opened its offensive with a massive artillery barrage against Israeli positions.[27] This barrage was accompanied by sporadic air strikes, as well as the insertion of SOF near the Israeli front line. None of these preliminary actions had any impact on the Israeli defenders: both the artillery barrage and the air strikes struck empty terrain, not Israeli positions, while the SOF were either killed or captured before they could inflict any damage or casualties. Moreover, these actions did not even provide cover for advancing Egyptian armored columns, as they were not coordinated with the movement of those columns.

Off to a poor start, the Egyptian offensive entered a steady downward spiral thereafter. The movement of the five armored thrusts, which was supposed to be coordinated in time and space, devolved throughout the morning into a disorganized series of isolated armored movements. In the 2nd (EG) Army Corps northern sector, the 15th (EG) Independent Armored Brigade, bolstered by infantry forces from the 18th (EG) Infantry Division, attacked the 11th and 204th (IS) Mechanized Infantry Brigades of Force Tiger in two waves. Israeli forces destroyed approximately 30 Egyptian tanks in these clashes before Egyptian forces called it a day. In the corps central sector, the 24th (EG) Armored Brigade (temporarily detached from the 23rd (EG) Mechanized Division), bolstered by infantry forces from the 2nd (EG) Infantry Division, attacked the 274th (IS) Armored Brigade of the 162nd (IS) Armored Division. With some assistance from the 500th (IS) Armored Brigade, which was backing it up, the 274th (IS) Armored Brigade repulsed two waves of Egyptian tanks and infantry, destroying approximately 40 tanks. And, in the corps southern sector, the 1st and 14th (EG) Armored Brigades of the 21st (EG) Armored Division, together with infantry forces from the 16th (EG) Infantry Division, attacked the 14th and 421st (IS) Armored Brigades of the 143rd (IS) Armored Division in two waves in a quest to break out toward Bir Gifgafah. In consequence of the combined defensive efforts of the two Israeli brigades, the 21st (EG) Armored Division lost on the order of 80–100 tanks in these thrusts.

The forces of the 3rd (EG) Army Corps suffered much the same fate. In the 3rd (EG) Army Corps northern sector, the 25th (EG) Independent Armored Brigade, supported by infantry forces from the 7th (EG) Infantry Division, twice advanced in the direction of the Gidi Pass only to be repelled by the 164th (IS) Armored Brigade of the

252nd (IS) Armored Division on both occasions, incurring the loss of approximately 20 tanks. In the corps central sector, the 3rd (EG) Armored Brigade (temporarily detached from the 4th (EG) Armored Division), along with infantry forces from the 19th (EG) Infantry Division, attempted to break through the Mitla Pass only to be halted in its tracks by a battalion of infantry from the 35th (IS) Parachute Infantry Brigade, supported by tanks and mechanized infantry under the command of the 401st (IS) Armored Brigade of the 252nd (IS) Armored Division. An IDF armored company commander would later remark of this action:

> I saw in the valley below many Egyptian tanks in a completely static position. We fired . . . at them from a very short distance. . . . It was over very quickly.[28]

An IDF armored battalion commander, swiftly deploying his armor once Egyptian tanks appeared on the horizon, would later comment on this same action, noting that: "We [got into] ideal firing positions. . . . [Egyptian armor] looked like a dream target [in] a turkey shoot."[29]

In the 3rd (EG) Army Corps southern sector, the 22nd (EG) Armored Brigade (temporarily detached from the 6th (EG) Mechanized Division), together with infantry forces from the 19th (EG) Infantry Division, advanced southward toward Ras Sudar only to be repelled by a scratch force of tanks and mechanized infantry under the command of the 252nd (IS) Armored Division. IAF fighter-bombers also played a significant role in repelling the 3rd (EG) Army Corps thrusts in the central and southern sectors by helping to stall and scatter the Egyptian columns.[30] Egyptian forces lost approximately 60 more tanks in the central and southern thrusts.

The Egyptian army's offensive, in sum, turned out to be an unmitigated disaster. According to some accounts of the battle, including accounts from Egyptian officers who participated in the Yom Kippur War, the army had lost approximately 265 tanks and 200 armored personnel carriers (APCs) in the fighting on October 14. Other accounts of the battle reach a somewhat more conservative estimate of 200–250 tanks, with the actual figure generally thought to be closer to the upper boundary. In either case, the Egyptian army had lost approximately one-third of the tanks it had committed to the offensive and approximately one-quarter of the tanks it had in the Sinai. Furthermore, the Egyptian army had failed to conquer any territory beyond its bridgeheads, and it had even lost a bit of ground in the area occupied by the 18th (EG) Infantry Division. The IDF, according to most accounts, had lost approximately 20 tanks—most of them from the 274th (IS) Armored Brigade—of which only six were complete write-offs, the remainder retrieved, repaired, and returned to service.

## Analysis of the Battle of October 14

An analysis of the battle in light of the six variables follows.

### *Operational Planning*

In stark contrast to its meticulous operational planning for the crossing of the Suez Canal prior to the Yom Kippur War, the Egyptian army's planning for its offensive to break out of its canal-side bridgeheads proved to be sorely deficient, apparently not going much beyond a quite vague outline of where and how its forces were to attack their IDF counterparts. The painstaking preparations necessary to synchronize five separate armored thrusts across two different army corps in order to put maximum simultaneous pressure on IDF forces were evidently absent, resulting in disjointed thrusts that were easily repulsed by Israeli defenders in serial fashion.

The IDF had a much easier task in respect of operational planning before the battle, because it was on defense during this engagement. Be that as it may, IDF planning for the battle nevertheless proved to be first-rate, with sufficient armored and mechanized infantry forces assembled and deployed in the proper positions to block each of the Egyptian army's armored thrusts, and with potent armored forces in reserve ready to respond immediately to any sector of the front where the Egyptian army threatened to break through Israeli defenses.

### *Intelligence and Reconnaissance*

During the Yom Kippur War, the Egyptian army obtained most of its information about the composition and location of Israeli forces in the Sinai from signals intelligence (SIGINT). IDF communications security often tended to be lax, particularly in the early days of hostilities, allowing the Egyptian army to intercept Israeli radio transmissions.[31] The primary Egyptian installation for doing so was located atop Jebel (Mount) Ataka. Early on October 14, Israeli SOF raided this installation, destroying its sophisticated Soviet-supplied electronic equipment.[32] Though the IDF mounted this special operation (SPEC

OP) in an effort to hamper the Egyptian army's capability to track its movements in connection with the anticipated Israeli counteroffensive across the Suez Canal, the loss of the Jebel Ataka installation undoubtedly sowed confusion among Egyptian forces on October 14, too, by eliminating their sole source of information about IDF defensive deployments. The Egyptian army, after all, had no human intelligence (HUMINT) sources within the IDF or the Meir government, nor did its forces generally engage in reconnaissance to map the composition and location of Israeli forces.

The IDF, in contrast, had extensive and accurate information about the capabilities and intentions of the Egyptian army on October 14. For starters, it possessed the Egyptian army's war plan, which a high-ranking official inside the Egyptian government had turned over to Israel before the Yom Kippur War.[33] This same official alerted the Israeli government to the imminent Egyptian offensive, while IDF SIGINT assets picked up and traced the movements of Egyptian armored forces across the Suez Canal.[34] Hence, the IDF had ample warning of the Egyptian offensive. Furthermore, IDF reconnaissance forces remained in constant visual contact with Egyptian forces on October 14, regularly passing information on the movement of those forces up the chain of command.[35] Israeli reconnaissance forces, as a matter of fact, transmitted invaluable information up the chain of command throughout the war. Indeed, as early as October 8, a reconnaissance force belonging to the 143rd (IS) Armored Division found a seam between the 2nd and 3rd (EG) Army Corps bridgeheads, a seam that the IDF would exploit to the fullest during its counteroffensive across the Suez Canal.[36]

## *Exploitation of Terrain*

The rugged and hilly terrain of the battlefield, particularly opposite the 3rd (EG) Army Corps bridgehead, favored Israeli forces, which held the high ground to the east of the Suez Canal.[37] Intimately familiar with the lay of the land, these forces chose optimal defensive positions from which to engage their Egyptian adversaries, positions that took full advantage of their long-range tank gunnery skills, while simultaneously minimizing their own exposure to return fire. Furthermore, Israeli tank crews were well versed in exploiting broken terrain to conceal their movements in order to strike at an adversary's flanks.

Egyptian forces, on the other hand, advanced out of the canal-side flatlands, which afforded those forces little cover. Moreover, Egyptian tank crews were not practiced in exploiting broken terrain to conceal their movements; therefore, they did not harness the rugged and hilly terrain to their own advantage in those situations in which they might have done so. Perhaps nowhere was the Israeli exploitation and Egyptian lack of exploitation of the terrain more telling on October 14 than in the advance of the 3rd (EG) Armored Brigade, which moved through a gulley in its effort to reach the Mitla Pass only to stumble into a three-sided Israeli ambush, suffering heavy losses before beating a retreat back to the 3rd (EG) Army Corps bridgehead.[38] Indeed, the only Israeli force to incur significant casualties in the fighting—the 274th (IS) Armored Brigade, which absorbed more than half of all Israeli losses in the battle—constituted the sole Israeli force on the front line not familiar with the terrain in its area of responsibility.[39]

### *Quantity and Quality of Forces*

In regard to quantity, an army on the offensive, according to a widely held view among military practitioners, requires at least a 3:1 advantage in fighting men and fighting machines in order to overcome an army on the defensive. While this rule of thumb has often turned out to be rather dubious—witness, for instance, the many successful IDF offensives in which its forces were actually outnumbered by their adversaries—it has usually been the case that an attacker requires significantly more men and machines than a defender to emerge victorious on the battlefield. On the eve of battle, the Egyptian army was nowhere close to possession of such a preponderance of fighting men and fighting machines. Its 1,000 tanks, for example, were not nearly enough to overcome 750 IDF tanks.

In regard to quality, the Egyptian military establishment has traditionally been very fond of citing the ostensible superiority of the Western arms in the IDF's possession over the Soviet arms in the Egyptian army's possession as a major cause of the latter's repeated defeats at the hands of the former in their wars with each other. A careful review of the relative strengths and weaknesses of Soviet and Western arms in the Egyptian and Israeli arsenals, respectively, however, reveals a much different pattern. In the Yom Kippur War, as in the earlier 1956 Sinai Campaign and the 1967 Six-Day War, the Egyptian army's Soviet-supplied arms were generally on a par with the IDF's Western-supplied arms.

The British and American tanks in the Israeli arsenal offered better firepower and, in the case of the former, better protection than their Soviet counterparts in the Egyptian arsenal. Soviet tanks, on the other hand, offered smaller targets and, at least in comparison to American tanks, were less likely to suffer from mechanical problems. Both Western and Soviet tanks, to put it another way, were superior in certain respects but inferior in others, so the overall qualitative difference between them, while perhaps slightly favoring the IDF, was not substantial. Two prominent American military analysts concluded that "prewar training levels and the independence of small units proved far more critical [in battle] in the long haul than tank technology or performance."[40] The same reasoning holds for APCs, a category in which perhaps the Egyptian army enjoyed a slim advantage. The Egyptian army had better antitank missiles and rockets than the IDF, but these weapons generally did not play a decisive role on the battlefield.[41] The quality of the artillery in both arsenals was roughly equivalent.[42] Finally, while the IAF had better fighter-bombers and munitions than the Egyptian air force, its capability to destroy tanks and APCs was not overly impressive.[43]

The real qualitative difference between the IDF and the Egyptian army rested not in their respective arsenals, but rather in their respective manpower. Israeli tank crews were far superior in general to their Egyptian counterparts, particularly in regard to their respective gunnery and maneuvering skills.[44] Israeli infantrymen and artillerymen were also superior to their Egyptian counterparts. Of equal—probably even greater—importance, the quality of IDF junior officers, from battalion commanders down to platoon commanders, far exceeded that of their Egyptian army counterparts in terms of both leadership ability and battlefield experience.

## *Command and Control*

Israeli command and control ($C^2$) at the operational level of warfare was often poor on the Sinai front during the Yom Kippur War, especially in the early days of hostilities, when IDF GHQ, IDF Southern Command HQ, and division commanders frequently issued perplexing—and sometimes contradictory—orders.[45] $C^2$ problems played a prominent role in the failure of the October 8 counterattack, a counterattack characterized by confusion and disorganization from the start. Nor did these problems disappear entirely later in the war, as the bickering among IDF GHQ, IDF Southern Command HQ, and division commanders reignited during the Israeli counteroffensive across the Suez Canal.[46] On October 14, however, such problems were absent, as IDF GHQ, IDF Southern Command HQ, and the division commanders all agreed on how best to tackle the Egyptian offensive. Moreover, a defensive battle, by its very nature, requires less in the way of $C^2$ mastery than an offensive battle, a fact that also worked in the IDF's favor on October 14.

If the $C^2$ of Israeli forces on the Sinai front at the operational level left a lot to be desired throughout much of the Yom Kippur War, the same definitely cannot be said of the $C^2$ of those forces at the tactical level of warfare, which was uniformly of a very high caliber. IDF junior officers routinely took the initiative and expertly led their forces. A United States Army officer would later comment about the IDF victory on the Golan front:

> Israeli commanders at the tactical level made many decisions at critical points in the fighting that turned the tide of battle [in] their favor. This was possible because leaders were leading from the front and had the authority to use their own initiative to make decisions without asking for permission (from above) to do so.[47]

This comment applies equally well to the contribution of IDF junior officers to the triumph in the Sinai on October 14.

Egyptian $C^2$ at the operational level during the early days of the Yom Kippur War constituted a mirror image of Israeli $C^2$. In part as a consequence of effective $C^2$, not only did the Egyptian army superbly orchestrate a large-scale crossing of the Suez Canal, but also it robustly defended its bridgeheads in the Sinai. Once it went over to the offensive on October 14, though, its $C^2$ at the operational level broke down completely. One observer very familiar with the history of the Egyptian army noted that:

> The Egyptian formations could not keep together and coordinate their operations. Thus, what should have been two divisional thrusts with three supporting attacks turned into [a series of] brigade-sized attacks scattered all across the front. The Egyptian columns got so spread out that they could not come to each other's aid as each came under attack from quick, maneuvering Israeli armor. Instead of two reinforced armored-division fists, the Israelis were hit almost randomly by piecemeal tank attacks that they pinned, outflanked, and crushed.[48]

In the same vein, an Israeli division commander noted that "instead of concentrated attacks, [Egyptian forces] tried to break through in brigade strength at [numerous] separate points along the front line."[49]

Egyptian $C^2$ at the tactical level of warfare was no more efficient.

Rather than maneuver their forces like their Israeli counterparts, Egyptian junior officers simply pushed their forces straight ahead into the waiting arms of the IDF, with no attempts made either to reconnoiter Israeli positions or to exploit the terrain to Egyptian advantage. As a result of this seemingly mindless approach, Egyptian forces all along the front line simply stumbled into well-prepared Israeli positions only to be thrown back with heavy losses.[50]

### *Combined Arms Warfare*

In response to a faulty assessment of why the IDF had been so successful in the Six-Day War, the Israeli defense establishment endorsed a nontraditional variant of CAW—one based on armor and aircraft at the expense of infantry, artillery, and engineers—in the years before the Yom Kippur War. A distinguished Israeli military analyst crisply summarized the problem as follows:

> The basic errors of the Israelis in the Yom Kippur War grew, paradoxically enough, out of their victory in the Six Day War. . . . Israeli commanders had emerged from the [Six-Day War] feeling that it was possible to accomplish everything with a tank and a plane and so built their armed forces in an unbalanced manner.[51]

The IDF's nontraditional approach to CAW, in tandem with its general unpreparedness for war, contributed greatly to the severe reverses suffered by Israeli forces in the Sinai during the early days of hostilities, as the Egyptian army had developed clever means for countering Israeli armor and airpower through the massive application of antitank and antiaircraft weapons, respectively.

Relying on its improvisational skills, the IDF quickly reverted to traditional CAW from October 9, integrating sufficient infantry and artillery into its armored forces. Its effort in this regard paid big dividends on October 14, when combined arms formations of tanks, infantry, and artillery worked hand in hand to smash every Egyptian army thrust. These IDF formations not only knocked out large numbers of Egyptian tanks, but also completely neutralized most of the Egyptian army's antitank infantry before it could even participate in the fighting. Moreover, Israeli airpower cooperated effectively with IDF ground forces to halt the Egyptian offensive in those areas in which Egyptian forces advanced beyond the range of their air defense umbrella.[52]

The Egyptian army's practice of CAW on October 14 turned out to be a mirror image of the IDF's practice. Whereas different arms of the Israeli forces fought as one, different arms of the Egyptian forces did not act in unison. The aforementioned observer of the Egyptian army declared that:

> [Egyptian] company, battalion, and brigade commanders failed to conduct proper combined-arms operations. For the most part, mechanized infantry went into battle close behind the tanks but never dismounted, and large numbers of troops were killed without making any contribution to the attack. . . . Egyptian artillery tried to lay down a covering barrage and also tried to provide fire support to units in trouble, but they [sic] simply could not.[53]

An Israeli tank battalion commander during the battle added that "[Egyptian] armor moved forward without air support, even where [Egyptian forces] lacked [an antiaircraft] umbrella."[54] The Egyptian army's abysmal practice of CAW on October 14 yielded a predictable outcome.

## Ramifications of the Battle of October 14

The Battle of October 14 has never been given a name, a fact that is emblematic of the relative obscurity to which it has been consigned by analysts of the war, certainly in comparison to the attention lavished upon the Egyptian crossing of the Suez Canal, as well as the subsequent Israeli counter-crossing of the waterway. Even some of the Israeli officers who participated in the battle tend to downplay the significance of this engagement. To wit, a deputy division commander and his staff assistant commented:

> [T]he defeat's impact on the [Egyptian] army's ability was negligible. At most, it was a slight dent in the fender. . . . [A]fter the attack failed, the Egyptians forfeited nothing in terms of war gains[,] except for losing manpower and equipment (which amounted to very minimal damage and may have been more of a nuisance than anything else).[55]

These particular officers contend that the battle's significance was confined to the psychological realm, serving to raise the IDF's morale and, perhaps, to lower that of its Egyptian adversary.

To go no further than the realm of morale, however, is to ignore the weighty military and diplomatic ramifications of the Battle of October 14, a battle that completely changed the complexion of the Yom Kippur War in the Sinai. Militarily speaking, from October 15 onward, the IDF went over to the offensive, the Egyptian army over to the defensive. The Battle of October 14, then, constituted a turning point in the war on the southern front. In a conversation with the Israeli prime minister, the commanding officer (CO) of IDF Southern Command captured the essence of the moment when he said that, "Our forces are themselves again and so are the Egyptians."[56] Moreover, the battle set the stage for another turning point on this front, the Israeli counteroffensive across the Suez Canal. Had Israeli forces not achieved a resounding victory on October 14, it is by no means clear that the Israeli government, which had the final say on the matter, would have authorized this counteroffensive, even though IDF GHQ had previously called for it regardless of what happened on that day.

Diplomatically speaking, the Israeli triumph on October 14 terminated the Israeli government's and IDF GHQ's interest in an early cease-fire. Prior to the battle, even proponents of a counteroffensive across the canal had envisioned it simply as a means to compel Egypt to accept a cease-fire. After the battle, the prevailing consensus inside both the Israeli government and the IDF, buoyed by a newfound sense of optimism, contended that the Israeli war aim on the southern front should now be to inflict a decisive defeat on the Egyptian army, to crush Egypt itself. To this end, once Israeli forces established a bridgehead across the canal, they swiftly advanced into the Egyptian interior, not only capturing a hefty chunk of territory there, but also eventually encircling the 3rd (EG) Army Corps. For its part, the Israeli government engaged in a diplomatic effort to delay a cease-fire for as long as possible in order to allow the IDF to accomplish its mission. Indeed, only the intervention of the superpowers, which brought hostilities to a close on October 24, saved the Egyptian army and Egypt itself from a catastrophic defeat that might well have exceeded in magnitude the one that they had suffered during the Six-Day War.

# 6

# Operation Mole Cricket 19

The primary objective of the Israel Air Force (IAF) during Arab–Israeli wars has been to establish air superiority over the battlefields, as well as over the State of Israel itself.[1] The pursuit of other objectives—assisting Israel Defense Forces (IDF) ground forces through close air support (CAS) and interdiction strikes, as well as undertaking strategic attacks inside enemy territory, for example—has taken a back seat to the pursuit of this objective.[2] The thinking of the IAF—and, for that matter, the entire IDF—has been that the air force must secure air superiority before it can attend in a determined manner to the pursuit of these other objectives. Command of the requisite airspace, in other words, has been considered a necessary (though not sufficient) condition for the effective pursuit of these other objectives.

The IAF, however, has not always had the luxury of acting in accordance with its war plans. The 1973 Yom Kippur War offers the foremost case in point. While the air force had developed elaborate plans to knock out the Egyptian and Syrian armies' integrated air defense systems (IADSs)—in the form of Operations Challenge 4 and Model 5, respectively—at the outset of battle, the Israeli government forbade an eleventh-hour first strike against these air defense arrays, fearing that the American government would not support Israel during the war if the latter were to land the initial blow.[3] Rather than tackle these IADSs—which had replaced the Egyptian and Syrian air forces as the major obstacle to Israeli air superiority by the end of the 1969–70 War of Attrition—in a systematic way according to its war plans, the IAF instead had no option but to attempt to overcome them early in the war in a haphazard fashion, during interims when it was not carrying out sustained CAS and/or interdiction strikes on behalf of very hard-pressed and heavily outnumbered IDF ground forces in the Sinai and on the Golan, which had been caught unawares by the Egyptian and Syrian army offensives.[4]

Challenge 4 and Model 5 were laid out on the drawing board as meticulous combined arms operations calling for the coordinated employment of fighter-bombers fitted out with a variety of precision-guided munitions (PGMs) and other ordnance, unmanned aerial vehicles (UAVs) fitted out as decoys, artillery batteries fitted out for especially long-range fire, and helicopters fitted out for electronic warfare (EW), based on accurate, up-to-date intelligence about the types and locations of Egyptian and Syrian surface-to-air missile (SAM) and antiaircraft artillery (AAA) batteries.[5] But both Challenge 4 and Model 5 unfolded very differently than envisioned before the war.

The IAF began to implement Challenge 4 during the morning of October 7, the second day of the Yom Kippur War, initially making some headway by attacking Egyptian air force bases and outlying AAA batteries in preparation for the main effort against the Egyptian IADS' SAM batteries; however, as a consequence of the perilous situation in which IDF forces on the Golan found themselves, IDF general headquarters (GHQ) abruptly canceled Challenge 4 in order to redirect Israeli airpower to the northern battlefield.[6] Following this decision, the IAF received the go ahead from IDF GHQ to initiate Model 5 later the same day. Both decisions turned out to be disastrous for the IAF. Not only did the former bring a complete halt to a potentially successful operation, but also the latter led to the most costly single operation of the war for the IAF. Attacking the Syrian IADS without accurate, up-to-date intelligence about the air defense layout (no preparatory reconnaissance sorties were flown), without EW support (the aforementioned helicopters had been deployed to the south in connection with Challenge 4), without UAV support (these drones were fired off far too early to serve as useful decoys), and without long-range artillery support (these batteries were not available at the time), IAF fighter-bombers suffered serious losses without inflicting notable damage to the Syrian array.[7]

The cancellation of Challenge 4 and the defeat of Model 5 had a substantial impact on IAF performance throughout the rest of the Yom Kippur War.[8] First, while the air force eventually knocked out a large part of the Egyptian IADS, opening up undefended flight corridors over the southern battlefield, which it then exploited on behalf of IDF ground forces, it do so only in the waning days of the war and only as a result of combined arms operations with ground forces that had crossed the Suez Canal into Egypt proper; hence, for much of the war, IAF CAS and interdiction strikes on behalf of IDF ground forces in the Sinai were either not forthcoming or of dubious effectiveness, in an effort to limit further fighter-bomber losses to the Egyptian air defense array, losses that had been high in the early days of the war. Second, the IAF was not able to punch similar holes in the Syrian IADS, though it did eventually compel the Syrian army to withdraw significant portions of the system from the front in order to defend vital targets in the country's hinterland. IAF strategic attacks deep inside Syria helped to cut down significantly on fighter-bomber losses over the Golan front later in the war as a result. But, while carrying out CAS and interdiction strikes on behalf of IDF ground forces during the early days of the war—CAS and interdiction strikes, incidentally, that were more effective than their counterparts on the Sinai front—the air force suffered very heavy losses.

In response to its less-than-decisive encounters with the Egyptian and Syrian IADSs in the Yom Kippur War, the IAF immediately went back to the drawing board in search of the tactics and technology with which to overcome an adversary's IADS in the next round of warfare. The deputy commander of the IAF during the Yom Kippur War, who would also go on to serve as the commanding officer (CO) of the air force during the 1982 Lebanon War, summarized the problem succinctly: "We have to find an answer to the SAM batteries."[9] Over the next decade, the IAF would incessantly refine the tactics and technology necessary to defeat these batteries, eventually resulting in the obliteration of the Syrian IADS in Lebanon, perhaps the most formidable antiaircraft defense array in the world in the early 1980s, during Operation Mole Cricket 19.

## Prelude to Operation Mole Cricket 19

The surprise attack launched by the Egyptian and Syrian armies at the outset of the Yom Kippur War, as well as the Israeli response to that attack, constituted a formidable obstacle to the implementation of the IAF's suppression of enemy air defenses (SEAD) campaign throughout the war.[10] But the attack and response did not constitute the sole obstacle. While the IAF entered the war with first-rate fighter-bombers, first-rate ordnance, and a sophisticated command and control ($C^2$) system for managing air battles, its capabilities in a couple of crucial areas were not sufficiently advanced to address the threat posed by enemy IADSs.[11] First, the IAF could not gather, process, and disseminate intelligence data in time for this information to be actionable, especially in respect of mobile SAM and AAA batteries. And, second, the IAF did not possess a dedicated $C^2$ system for air–ground attack, including SEAD, which added to the air force's woes after its intricate prewar operations were not executed as planned, compelling it to implement hastily developed alternative operations while hostilities were in progress.

The dismal results yielded by Challenge 4 and Model 5 notwithstanding, the IAF remained fully committed to the concept of a large-scale SEAD operation at the outset of the next war as the best means to gain and hold air superiority over the battlefield. The IAF operations directorate rejected the concept of a series of limited, discrete SEAD operations designed to knock out an enemy's IADS one segment at a time—a concept that had essentially guided IAF SEAD operations on the Sinai front during the waning days of the Yom Kippur War. The IAF CO during the Lebanon War artfully described the air force's post-Yom Kippur War concept in respect of SEAD when he stated that it favored a symphony over solos.[12]

In the decade between the Yom Kippur War and the Lebanon War, the IAF methodically assembled all of the instruments that it required to play its SEAD symphony.[13] These instruments included: advanced fighter-bombers equipped with state-of-the art PGMs, avionics systems, and EW systems; airborne electronic warfare (AEW) aircraft equipped with state-of-the-art jamming systems; airborne warning and control system (AWACS) aircraft equipped with state-of-the-art command, control, communications, and computer systems; and a number of types of state-of-the-art UAV to serve in either the reconnaissance/intelligence role or in the decoy role. In addition to these air force assets, the IAF also had at its disposal IDF Artillery Corps long-range guns and rockets, as well as short-range surface-to-surface (SSM) missiles. Moreover, not only did the IAF assemble all of these assets, but also it integrated them into a seamless web under the aegis of a highly sophisticated command, control, communications, and intelligence ($C^3I$) system.

All of this technology, however, might well have been for naught had it not been for the rigorous and realistic training effort undertaken by the IAF in respect of SEAD.[14] Aircrews routinely rehearsed attacks against simulated SAM and AAA batteries on training ranges located in Israel's Negev Desert, while air force support personnel practiced the jamming of enemy communications and radar systems. Furthermore, the IAF and IDF Artillery Corps developed combined arms tactics in order to synchronize to optimal advantage the employment of air strikes and artillery bombardments against an enemy IADS. Two military analysts summed up the Israeli effort as follows:

> [Operation Mole Cricket 19] was backed [by] a solid and self-critical training and evaluation effort. . . .[The IAF] combined dry

> runs and mission mock-ups with practical experience over Lebanon . . ., [resulting in] a performance training-and-evaluation effort that [American] air power has never matched. . . . This training-and-evaluation effort was also the key to a C3I sequence to allow the coordination of air and artillery strikes under [SEAD] conditions.[15]

While the IAF patiently prepared itself for a large-scale SEAD operation in the decade after the Yom Kippur War, the Israel–Lebanon border grew increasingly "hot." Indeed, this border remained Israel's only active front, with the IDF and the Palestine Liberation Organization (PLO) regularly trading blows across the frontier over the years.[16] The triggers behind these flare-ups resided in a string of murderous terrorist attacks against Israeli civilians carried out by one or another of the PLO's various factions.[17] In the wake of a particularly gruesome attack that killed dozens of civilians, the Israeli government authorized the IDF to undertake Operation Litani, a limited incursion into Lebanon intended to clear PLO forces south of the Litani River. Though the IDF dealt a hard blow to the PLO during the fighting, the latter still possessed significant forces throughout Lebanon after the completion of the operation; therefore, renewed cross-border skirmishing was not long in coming, setting the stage for a war between Israel and its Syrian and Palestinian adversaries.

## Operation Mole Cricket 19

The origins of the operation lay in the Syrian government's decision in the spring of 1981 to begin to erect an IADS in the Bekaa in order to contest Israeli air superiority over Lebanon, a decision taken in response to the downing by the IAF of two troop-transport helicopters participating in Syrian army assaults against Israeli-supported Lebanese Maronite Christian forces in the eastern sector of the country.[18] The Israeli government appeared ready to authorize the IAF to destroy this incipient IADS immediately upon its deployment, but American diplomatic intervention prevented an escalation of hostilities between Israel and Syria—however, without addressing the fundamental conflict between the countries vis-à-vis the extensive and expanding Syrian military presence in Lebanon, the pretext for which was the need to stamp out the civil war then engulfing that country. Similarly, American diplomatic intervention prevented an escalation of hostilities between Israel and the PLO in the summer of 1981 after yet another bout of cross-border skirmishing—once more, however, without resolving the underlying conflict between the parties. Relative quiet would prevail in Lebanon for the better part of a year as a consequence of these American diplomatic interventions.

Still, by the spring of 1982, full-scale war between Israel and the PLO had become inevitable.[19] The attempted assassination of the Israeli ambassador to Great Britain by one of the factions of the PLO served as the match that set off the conflagration. Operation Peace for Galilee, the Israeli invasion of Lebanon, began on June 6. At midday on June 9, the fourth day of the war, the Israeli government gave its final approval for the IAF to execute Mole Cricket 19. While Peace for Galilee had originally been drawn up as a limited incursion designed simply to drive PLO forces from southern Lebanon, it was now about to morph into a full-scale war with both PLO and Syrian forces.[20]

Mole Cricket 19 unfolded in a series of waves.[21] The first wave—and the one most crucial for success—saw air- and ground-launched decoy drones fitted with special electronic emitters to make them appear to be fighter-bombers on Syrian radar screens wing their way into the Bekaa. Syrian air defense crews, as predicted by the IAF, immediately swallowed the bait. They activated their IADS' radar network and began to fire SAMs at these targets. Concomitantly, IAF fighter-bombers flying at low altitude to mask their approach from this radar network and fitted with antiradiation missiles (ARMs), munitions optimized to knock out radar systems, as well as other types of PGM optimized to attack $C^2$ systems, popped up entirely unexpectedly over the Bekaa and swiftly launched their ordnance, which homed in on and destroyed almost all of the components of the radar and $C^2$ networks. Firing in synchronization with this air strike, the long-range guns, long-range rockets, and short-range antiradiation SSMs of the IDF Artillery Corps added to the destruction.

The IAF's $C^3$I system, which relied heavily on real-time intelligence data provided by reconnaissance UAVs that were over the Bekaa throughout the operation, coordinated these different elements of the first wave flawlessly. Furthermore, AEW aircraft, EW helicopters, EW ground stations, as well as EW pods on the fighter-bombers themselves, ensured that Syrian fire-control systems were very effectively jammed, so much so that many SAMs were observed to veer wildly off course, exploding harmlessly far away from any target.

Mole Cricket 19's second wave followed shortly after the first, taking full advantage of the fact that the IAF had essentially blinded the Syrian IADS. Fighter-bombers fitted with PGMs, general purpose (GP) bombs, and cluster bomb units (CBUs), the last a weapon that releases many small bomblets over a designated area, now went after the "nuts and bolts" of the SAM and AAA batteries themselves, smashing vehicles, missiles, guns, and other assorted hardware, as well as killing the crews manning them. Though the first and second waves of Mole Cricket 19 had more or less accomplished the destruction of the Syrian army's IADS in the Bekaa, the IAF launched another two waves of fighter-bombers against the air defense array to clean up the remnants.

The results of the operation were nothing short of spectacular. The IAF had either utterly destroyed or badly damaged 17 of the 19 SAM batteries in the Bekaa, as well as several others located inside Syria proper.[22] Indeed, by the end of the Lebanon War, the IAF would knock out a total of 29 SAM batteries, including a number on June 10 comprised of the two surviving batteries and a few additional batteries deployed to the Bekaa overnight as reinforcements.[23] The air force wreaked a similar level of havoc among Syrian AAA batteries during Mole Cricket 19. Moreover, the IAF inflicted all of this destruction without the loss of a single fighter-bomber and, reportedly, with only 125 SEAD sorties.[24] The entire operation lasted only a couple of hours, with the most crippling damage—the destruction of the IADS' radar and $C^2$ networks—occurring in the first 10–20 minutes.

To make certain that the Syrian air force did not interfere with its SEAD effort, the IAF mounted a major counter-air effort against the former during Mole Cricket 19.[25] Initially, the Syrian air force made no attempt to intercept Israeli fighter-bombers, as Syrian army GHQ thought that the IADS could defend itself adequately against the IAF; however, once it became evident that the IADS could not do so, Syrian air force fighter-bombers began to head to the Bekaa like moths to a flame. IAF UAVs and AWACS aircraft tracked the Syrian fighter-bombers continuously as soon as the latter moved to takeoff positions on their runways. Simultaneously, IAF AEW aircraft and ground stations jammed the communications links between Syrian air force ground control intercept (GCI) stations and the fighter-bombers, leaving Syrian pilots in the dark, as they depended heavily on GCI to put them into position to confront their Israeli counterparts. In stark contrast, IAF AEW aircraft placed Israeli fighter-bombers in an optimal position from which to engage their Syrian counterparts. Consequently, on June 9, IAF fighter-bombers shot down more than 20 Syrian aircraft with no loss to themselves during some 80 counter-air sorties in the largest single air-to-air battle since the Second World War.[26] This pattern repeated itself on June 10 and 11, with approximately 40 more Syrian aircraft falling in air-to-air battles, again with no loss to the IAF.[27] By the end of the Lebanon War, IAF fighter-bombers had shot down more than 90 Syrian aircraft at no cost to themselves.[28]

## Analysis of Operation Mole Cricket 19

An analysis of the operation in light of the six variables follows.

### *Operational Planning*

IAF planning for Mole Cricket 19 may only be described as meticulous in every respect. So much so that the aforementioned military analysts concluded that:

> Maps . . . [showed] the exact location, effective range, and sensor coverage of . . . every radar [and SAM battery] in or affecting the Bekaa target area. . . . [The IAF] had the equivalent of a high resolution photomosaic of the Bekaa when it attacked[,] which showed the layout of every major individual Syrian air-defense site. . . . [The IAF] also had the equivalent of an operations map for each [SAM and AAA] site that showed its electromagnetic characteristics, the required active and passive counterforce options, the proper method of air attack, and precise artillery coordinates. . . .[B]oth the precise target registration and proper angle of air attack were confirmed [ahead of time].[29]

They went on to add that:

> The command and control phase of the attack plan was detailed to the point where a specific choice of the best attack mix for each [SAM and AAA] site could be updated on the day of the attack, with choices [among] conventional iron bombs, cluster munitions, [various types of air-launched PGM], [and] the use of given numbers and types of . . . artillery rounds.[30]

And, finally, they pointed out that:

> [The IAF's] strike [plan was] backed by a general active and passive [electronic] countermeasures plan, along with a full air-cover plan. [The IAF's] use of [reconnaissance] aircraft and [UAVs] to confirm damage and [to] help [inform] follow-up attacks was

> equally well planned, as [were] reserve [and] alternate attack [capabilities].[31]

The IAF, in other words, knew the answer to every what, where, when, and how question in connection with every target on its list. The air force's manifest desire for an aerial symphony in respect of SEAD, in short, was fully enshrined in the planning of Mole Cricket 19. Likewise, the IAF carefully planned for the air-to-air engagements that it anticipated would accompany the execution of the operation.

The Syrian army, on the other hand, blatantly neglected to plan properly for an Israeli aerial onslaught in the Bekaa.[32] It mindlessly followed standard Soviet practice in deploying its IADS (with fewer search radars than called for under Soviet air defense doctrine), and it rigidly trained its air defense crews (and support personnel) in standard Soviet procedures and tactics, with apparently no consideration given to the IAF's technology and tactics. When the day of reckoning arrived for the Syrian army in the Bekaa, therefore, it IADS was wholly unready to meet the test.

### *Intelligence and Reconnaissance*

The delay in the outbreak of the Lebanon War turned out to be a blessing in respect of Mole Cricket 19, because it gave Israel time to build a comprehensive intelligence picture of the Syrian air defense layout in the country.[33] In the months between the introduction of Syrian air defenses into the Bekaa and the execution of Mole Cricket 19, Israel's various intelligence assets systematically collected information on every facet of the Syrian army's IADS via a combination of PHOTINT (photographic intelligence), SIGINT (signals intelligence), and HUMINT (human intelligence). The IAF knew the precise territorial layout of the Syrian army's IADS at any given moment as a consequence of regular photoreconnaissance sorties over the Bekaa by both manned aircraft and UAVs. At the cost of a few expendable decoy drones lost to antiaircraft fire, the IAF acquired equally precise information on the electronic footprints of the SAM batteries, as aircraft and ground stations recorded operating frequencies when Syrian air defense crews activated those batteries in attempts to shoot down the Israeli drones. IDF special operations forces (SOF)—and, perhaps, civilian agents—gathered additional data on the disposition of the Syrian army's IADS through ground reconnaissance. On the day when the IAF initiated Mole Cricket 19, in short, the IAF had a complete portrait of the Syrian IADS, including a firm sense of the competence of the SAM crews manning the 19 batteries deployed in the air defense array. And, throughout the operation itself, the IAF had in its hands continuous real-time, actionable intelligence information courtesy of its extensive PHOTINT and SIGINT assets.

The Syrian army's intelligence and reconnaissance efforts before and during the Lebanon War constituted a mirror image of the Israeli efforts. Indeed, one military analyst intimately familiar with Syrian military performance in the Arab–Israeli wars concluded that:

> Syrian military intelligence performed dismally. . . . [It] completely dropped the ball with regard to the development of Israeli military forces and operations. [It was] entirely unaware of some of the systems the Israelis employed to destroy the SAM systems in the Bekaa and did not even foresee that the Israelis would make a major effort to neutralize Syria's ground-based air defenses.[34]

### *Exploitation of Terrain*

The IAF took advantage of the rugged terrain in the Bekaa region during Mole Cricket 19. Its fighter-bombers, especially those in the first wave, following carefully pre-planned nap-of-the-earth (NOE) flight routes, approached their targets at very low altitude, staying out of sight of Syrian radar screens by "hiding" behind the mountains in the area, suddenly popping up seemingly from nowhere to deliver their ordnance before the Syrian IADS could react meaningfully to their presence, while decoy UAVs approached the IADS by more conventional flight routes in order to draw the Syrian army's attention.[35]

For its part, the Syrian army clearly did not make optimal use of the terrain when it established its IADS in the Bekaa. The army neglected to consider how the mountainous terrain of the region would affect its radar systems' capability to detect low-flying aircraft.[36] Nor did it take into account potential wind conditions in the area. On June 9, these conditions were such that, when the Syrian army tried to generate a smokescreen in an attempt to hide the locations of its SAM and AAA batteries from the IAF, discrete plumes of smoke hung in the air above individual batteries instead of dispersing into one large, amorphous screen, providing excellent visual cues to the attacking fighter-bombers.[37]

### *Quantity and Quality of Forces*

The IAF assembled a superb order of battle (ORBAT) on the eve of Mole Cricket 19.[38] Quantitatively speaking, it had more than enough fighter-bombers, UAVs, AEW and AWACS aircraft, attached IDF artillery systems, and so on to execute a major SEAD campaign, as well as more than enough fighter-bombers and ancillary assets to contend with the Syrian air force in air-to-air combat. Indeed, most of the IAF's ORBAT had been assembled specifically with the SEAD and air-to-air combat missions in mind. And, qualitatively speaking, it had put together an ORBAT consisting of the most sophisticated equipment in the world operated by the most highly trained and the most experienced aircrews in the world.

The densely laid out SAM batteries deployed in the Bekaa as part of the Syrian army's IADS consisted of an advanced Soviet model.[39] Likewise, the AAA batteries consisted of advanced Soviet models. Relatively modern shoulder-fired SAMs and relatively modern Syrian air force interceptors were also available in some numbers to support the Syrian army's air defense array. The Syrian operators and their Soviet advisors believed—incorrectly—that the $C^2$ system linking the various components of the IADS to each other, as well as linking the IADS to the Syrian army's GHQ, to be robust. In the days prior to the IAF assault, therefore, both the Syrian and Soviet army GHQs thought—again, incorrectly—that the IADS in the Bekaa would be able to withstand a possible Israeli attack.

### *Command and Control*

Fully aware of the importance of battle management, the IAF systematically built the most sophisticated $C^3I$ system in the world for both SEAD and counter-air operations during the years between the Yom Kippur War and the Lebanon War, a unique system at the time in that it incorporated assets—particularly UAVs—capable of providing real-time intelligence data not only to the IAF's command post at IDF GHQ, but also to IAF assets in the field. One airpower analyst reported that:

> The IAF command post . . . gave [the air force commander] a real-time . . . picture of [Mole Cricket 19] through various data links. [AWACS aircraft] with their airborne surveillance radar[s] down-linked their pictures to the command post. [UAVs] provided video[, which] helped [to] provide [the] locations of the Syrian

> SAM batteries [continuously throughout the operation]. . . . The IAF also set up two-way voice communications between [the IAF commander] and his pilots. This real-time command, control, and intelligence capability . . . delivered what [the IAF commander] termed the "real-time intelligence" to . . . the strike force.[40]

In respect of the IAF counter-air effort during Mole Cricket 19, another airpower analyst commented that:

> Israeli [UAVs] used [their] electro-optical zoom lense[s] and digital data link[s] to provide real-time imagery of Syrian [fighter-bombers] positioned for takeoff. [AWACS aircraft] . . . [were then] able to pick up the [Syrian aircraft] on radar as soon as they left their runways and relayed intercept vectors to IAF [fighter-bombers].[41]

The capabilities of the Israeli $C^3I$ system, in short, contributed significantly to the IAF's success in Mole Cricket 19 and the accompanying counter-air effort.

While the Israeli $C^3I$ system helped to ensure the Israeli victory, the Syrian $C^2$ system just as surely helped to ensure the Syrian defeat. The Syrian army and air force did not take the issue of $C^2$ seriously in the years between the Yom Kippur War and the Lebanon War. According to two military analysts:

> Syria organized its air force and [IADS] into highly compartmented command [and control] structures [that] were not trained to cooperate and, moreover, lacked the software and hardware for effective communication.[42]

Hence, when the IAF commenced Mole Cricket 19, another analyst observed, the Syrian IADS':

> [SAM] batteries . . . had difficulty communicating with each other and with the . . . AAA [batteries] supposed to defend them[; consequently, the SAM batteries] could not coordinate their responses to the Israeli attack.[43]

Furthermore, this analyst continued, when Syrian fighter-bombers rose to defend their air defense array in the Bekaa, the IAF:

> began a comprehensive electronic-jamming campaign that sev-

> ered the communications links between the Syrian [aircraft] and their GCI sites. Deprived of GCI guidance, [Syrian fighter-bomber] . . . formations immediately dissolved . . .[44]

Flying without ground communications and ground radar support, Syrian aircraft fell easy prey to IAF combat air patrols (CAPs) directed to their targets by AWACS aircraft. Many Syrian fighter-bombers shot down by the IAF were apparently taken completely by surprise by Israeli fighter-bombers engaging them from unexpected directions.

### *Combined Arms Warfare*

The IAF learned the lesson of the importance of combined arms warfare (CAW) the hard way during the Yom Kippur War. While Model 5 had clearly been planned as a CAW operation, it just as clearly did not unfold that way, as fighter-bombers pressed home their attack without the benefit of support from decoy UAVs, EW helicopters, or long-range artillery fire. In its postmortem of its SEAD efforts during the Yom Kippur War, the IAF laid much of the blame for the Model 5 debacle on the lack of CAW during the operation. Consequently, it committed itself even more firmly than ever before to the practice of CAW during any future SEAD operations.

Naturally, then, the IAF planned Mole Cricket 19 as an intricate CAW operation. Under the umbrella of its $C^3I$ system, the air force tightly integrated the deployment of fighter-bombers, UAVs, AEW aircraft, EW helicopters, EW ground stations, as well as long-range artillery and short-range SSMs. Likewise, during the accompanying counter-air effort, the IAF closely integrated the deployment of fighter-bombers, UAVs, AEW aircraft, and AWACS aircraft. The sequential and/or simultaneous deployment of all of these complementary assets contributed significantly to the IAF's lopsided victories in both SEAD and counter-air actions.

So did the Syrian army and air force's failure to practice CAW at the most rudimentary level. Not only were the different elements of the army's IADS in the Bekaa unable to coordinate their actions in the face of the Israeli aerial blitz, but also the air force's fighter-bombers were unable to coordinate their actions with the actions of the army's IADS. In respect of the latter, one military analyst commented that:

> [Syrian forces were] incapable of coordinating the operations of [aircraft] and ground-based air defense units; the result was that

> [the two] could only act either at different times or in different areas, but they could not integrate their efforts.[45]

The Syrian air force, as a matter of fact, did not even attempt to disrupt the IAF's attack on the IADS until it was too late. The air defense array had essentially been knocked out before the first Syrian fighter-bombers engaged their Israeli counterparts over the Bekaa.

## Ramifications of Operation Mole Cricket 19

In consequence of the outstanding success of Mole Cricket 19, along with the equally outstanding success of the associated counter-air effort, the IAF established air superiority over Lebanon early in the war and maintained it throughout the fighting. Air superiority, in turn, allowed the IAF to carry out more effective CAS and interdiction strikes than it had during the Yom Kippur War.[46] On June 10, to highlight just one example, IAF fighter-bombers caught a Syrian armored brigade in the open as it moved into the Bekaa, destroying 20 tanks and many other armored vehicles.[47] IAF antitank helicopters, free of the threat of Syrian fighter-bombers, seem to have been particularly effective in the CAS role.[48] While Mole Cricket 19 certainly contributed positively to the greater Israeli war effort, including the successful drive by IDF ground forces to the outskirts of Beirut, the largely mountainous and urban terrain of Lebanon, it must be said, did limit to some extent the overall impact of IAF CAS and interdiction strikes.[49] Mountainous terrain often provided Syrian ground forces with very good concealment against CAS and interdiction strikes, while urban terrain often provided both Syrian and PLO forces with equally good concealment, especially in light of the IDF's efforts to minimize collateral damage to surrounding civilian populations and property.

Mole Cricket 19, parenthetically, had a significant effect on strategic affairs unconnected to the Lebanon War. First, the operation served as an eye-opening test bed for the current Revolution in Military Affairs (RMA).[50] An American defense official captured the essence of an RMA by defining it as:

> . . . a major change in the nature of warfare brought about by the innovative application of new technologies which, combined with dramatic changes in military doctrine and operational and organizational concepts, fundamentally alters the character and conduct of military operations.[51]

Mole Cricket 19, asserted a military analyst, demonstrated the viability of applying such new technologies, as well as such new operational and organizational concepts, to the battlefield in order to bring about sweeping change to the way in which wars would be fought in the future by:

> [tailoring] intelligence and target acquisition systems, command and control elements, and precision stand-off fire into one integrated system and uninterruptedly [implementing] real-time sensor-to-shooter loops, for a relatively prolonged period of time.[52]

The United States' armed forces would extend the scope and sophistication of this RMA during the 1991 Gulf War and then in such places as Afghanistan and Iraq in the new millenium.[53] One airpower analyst put it thusly:

> [Operation Mole Cricket 19] . . . provided a preview of the technological marvels of the . . . Gulf War, with the US Air Force's destruction of integrated air defenses, to increasing real-time control by those in charge of fighter operations, to [l]aser-[g]uided [b]ombs hitting tanks in the desert.[54]

And, second, Mole Cricket 19 altered the balance of power between North Atlantic Treaty Organization (NATO) and Warsaw Pact (WP) forces. In consequence of events during the Lebanon War, Soviet army GHQ realized that Soviet military technology had fallen well behind Western military technology, sending shock waves through the halls of power in the Soviet government. A Czechoslovakian general stationed in the Soviet Union at the time of the war would later go so far as to contend that Mole Cricket 19 was a conspicuous part of a train of events that culminated in the dissolution of the Soviet Union, as well as in the end of the Cold War, less than a decade later, a view that also has proponents within post-Soviet Russia.[55] The dissolution of the Soviet Union, of course, was also of great strategic benefit to Israel.

# 7

# Operations Hell and Spring of Youth

The Israeli defense establishment has traditionally placed the various national security challenges faced by the State of Israel into two broad categories: "basic" security challenges and "current" security challenges.[1] The defense establishment has defined as basic security challenges the ones that put at risk the fundamental integrity of the country. This category includes war with adversarial states and attack with weapons of mass destruction (WMD). The defense establishment has defined as current security challenges the ones that, while they do note pose an existential threat to the country, put at risk its citizenry. This category includes such lower-level challenges as insurgency, terrorism, and border skirmishing.

With the armies of the Arab confrontation states licking their wounds, current security challenges overshadowed basic security challenges in the immediate aftermath of the 1967 Six-Day War. The country's national security situation resembled the ones that had prevailed in the years between the 1947–49 War of Independence and the 1956 Sinai Campaign, as well in the years just prior to the Six-Day War.[2] Like the fighting in these earlier rounds, Arab nonstate insurgent/terrorist organizations engaged in insurgency and terrorism against Israel, frequently triggering border skirmishes between it and the Arab confrontation states, though the scope and intensity of hostilities were now considerably greater than in the earlier rounds.

To defend the country during this new round of fighting, the Israel Defense Forces (IDF) resorted to essentially the same combination of defensive and offensive means as it had during the earlier rounds. The defensive measures included electronic surveillance, mobile infantry patrols, border checkpoints and observation posts, minefields, and fencing/barbed wire, all weaved together in a flexible web, permitting a swift response to an insurgent or terrorist attack. The offensive measures included targeted killings of insurgent/terrorist commanders and raids inside "host" states in which the insurgent/terrorist organizations had established bases.

## Prelude to Operation Hell

Established by the Arab states in 1964—ironically, in an effort on their part to delay going to war with Israel—the Palestine Liberation Organization (PLO) waged an insurgency/terrorist campaign against the "Zionist entity" in the years before the Six-Day War.[3] Fanatically committed to the total destruction of its adversary, the PLO carried out over 100 attacks against Israel in the two-and-a-half years prior to the Six Day War; however, as a result of the rather amateurish nature of these attacks, this insurgency/terrorist campaign inflicted few casualties and caused little damage. The PLO, in other words, played only a very minor role in the Arab–Israeli conflict before the Six-Day War.

The outcome of this war, though, would pave the way for the organization's swift rise to prominence. First, the crushing defeat suffered by the Arab armies had not only taken them largely (albeit temporarily) out of the fight against Israel, but also it had largely discredited them in the eyes of the Arab world as the standard-bearers of the fight against Israel. The PLO, which had not been discredited by the outcome of the war, immediately assumed this mantle, its ranks quickly swelling with recruits and its arsenal quickly filling with arms. Second, Israel's conquest of Judea, Samaria, and Gaza during the war—all three territories containing substantial Palestinian populations—appeared to offer the PLO fertile ground over which to conduct an insurgency/terrorist campaign.

Predictably, then, the PLO renewed its insurgency/terrorist campaign against Israel shortly after the Six-Day War.[4] It waged this campaign both inside and outside of Judea, Samaria, and Gaza. Inside these territories, the PLO campaign never amounted to much. The Palestinian population of Judea, Samaria, and Gaza may not have harbored warm feelings for Israel, but a concerted Israeli effort to improve socioeconomic and health conditions in these territories, coupled with a general disinterest among resident Palestinians in "armed resistance," ensured that the insurgency/terrorist campaign did not gain traction there. Moreover, IDF forces were able to extinguish any incipient sparks of resistance in Judea, Samaria, and Gaza through a combination of defensive and offensive measures, such as vigorous intelligence collection, regular mobile infantry patrols, and occasional targeted killings of insurgent/terrorist operatives.

Outside of Judea, Samaria, and Gaza, the PLO campaign constituted a more ominous development for Israel. In the wake of the Six-Day War, the PLO had set up a network of bases in the Hashemite Kingdom of Jordan, many of these bases in close proximity to the Jordan River. From these bases, it routinely dispatched insurgent/terrorist squads to conduct attacks inside Israel and the "occupied territories." While relatively few of these squads proved able to penetrate Israeli defenses in order actually to carry out attacks, IDF forces suffered significant casualties in preventing the squads from reaching their objectives. IDF general headquarters (GHQ), therefore, soon decided upon a more proactive response to the predicament brought about by border infiltration along Israel's central frontier.

## Operation Hell

Operation Hell, also known as the Battle of Karameh, constituted one of the fruits of this more proactive response to the PLO insurgency/terrorist campaign against Israel.[5] IDF GHQ set for its forces the primary objective of destroying the large PLO base at Karameh in order to drive the latter's forces out of the Jordan Valley and away from the Israeli border.[6] IDF GHQ possibly also set for its forces the secondary objective of delivering a hard blow to the Jordanian army, the Arab Legion, in order to compel Jordan to crack down on PLO forces on its territory.

Whatever the truth of the matter, IDF GHQ expected that the Arab Legion would intervene on a large scale during a major Israeli operation on Jordanian territory; therefore, IDF GHQ assembled an impressive order of battle (ORBAT) for Hell, which included elements of the 7th and 60th (IS) Armored Brigades, elements of the 80th (IS) Infantry Brigade, and much of the 35th (IS) Parachute Infantry Brigade (consisting of elements of the 50th and 202nd (IS) Parachute Infantry Battalions, along with elements of Sayeret Tzanchanim, the brigade's reconnaissance unit), as well as elements of an engineer battalion.[7] These forces would be supported by artillery and airpower as necessary. For its part, the Arab Legion, expecting a major Israeli incursion into Jordanian territory, also assembled an impressive ORBAT for the upcoming battle, which included the 1st (AL) Infantry Division (with three infantry brigades), the 60th (AL) Armored Brigade, and attached artillery and antitank forces.[8] Hundreds of PLO insurgents/terrorists in and around Karameh, who were equipped with small arms and antitank weapons, added to the Arab ORBAT.

During the early morning of March 21, 1968, the IDF commenced the operation, launching a four-pronged assault on the PLO base at Karameh. Israel Air Force (IAF) helicopters transported elements of Sayeret Tzanchanim to points east of Karameh in order to cut off any PLO forces attempting to escape the area. Concomitantly, three mixed armored–infantry columns crossed the Jordan River with the intention of converging on Karameh from different directions. The main column, comprised of elements of the 7th (IS) Armored Brigade, as well as elements of the 50th and 202nd (IS) Parachute Infantry Battalions, penetrated Arab Legion defenses along the Jordan River, split up into several smaller columns, all of which pushed on toward Karameh, joining up with Sayeret Tzanchanim forces, which were already engaged against the combined Arab Legion–PLO forces there. Together, these Israeli forces thoroughly destroyed the PLO base and inflicted heavy casualties on combined Arab Legion–PLO forces, though at some cost to themselves.

The two other columns—one to the north and one to the south of the main column—did not fare nearly as well. The northern column crossed the river and created a solid bridgehead on Jordanian territory, but it soon got bogged down in front of Arab Legion defenses in the area, which blocked the column's advance. Israeli artillery fire and air strikes in support of this column proved to be unable to inflict substantial damage on the heavily bunkered Arab Legion positions located on the high ground to the east of the Jordan Valley. The southern column did not even establish a firm bridgehead across the river. Like its northern counterpart, this column was stymied by heavily bunkered Arab Legion defenses located on the high ground to the east. Israeli artillery fire and air strikes in support of this column also proved to be ineffective against Arab Legion defenses in the area. Nevertheless, these columns gave as good as they received from the Arab Legion in terms of losses among fighting men and fighting machines.

After a day of battle, the IDF withdrew all of its forces back across the Jordan River in an orderly fashion, reckoning that any further fighting would result in additional costs that would outweigh any additional benefits. Jordanian and Palestinian propaganda later cited the Israeli withdrawal as "proof" that they had won the Battle of Karameh.[9] In reality, however, the IDF had won the battle, though perhaps not as decisively as it had anticipated beforehand.

IDF forces had, after all, accomplished the primary objective of the operation, the destruction of the PLO base at Karameh. Still, they had suffered considerable casualties in the process. The IDF lost about 30 soldiers, as well as about 10 tanks and other armored vehicles in the fighting.[10] The IAF also lost one fighter-bomber to Arab Legion air defenses. Arab Legion and PLO losses, on the other hand, were much greater. The Arab Legion lost about 60 soldiers, as well as about 30 tanks and other armored vehicles, while the PLO lost 100–150 insurgents/terrorists (with another 125–150 taken prisoner). Relative losses become even more lopsided if the approximately 75 Arab Legion soldiers and PLO insurgents/terrorists killed during Operation Remedy—a simultaneous diversionary raid carried out by a combined IDF armor–parachute infantry task force against PLO installations south of the Dead Sea area—are added to the total.[11] The IDF suffered no losses during this diversionary operation.

## Analysis of Operation Hell

An analysis of the operation in light of the six variables follows.

### *Operational Planning*

The IDF devised a feasible and creative operational plan to "swarm" the PLO base at Karameh from every direction in order to confuse and overwhelm the installation's defenders, and it assembled the forces needed to accomplish this objective; however, the plan's one serious flaw—its underestimation of the Arab Legion's eagerness to contest a sizable IDF incursion into Jordanian territory, despite an awareness that the Jordanian forces would certainly be inclined to fight—resulted in a less decisive outcome of the battle than forecast beforehand.

For its part, the Arab Legion devised a sensible operational plan to defend Karameh, and it arranged its forces properly in and around the Jordan River Valley in order to do so. "[Arab Legion] forces," according to a prominent military historian very familiar with Arab armies, "were well sited to block all major axes of advance, to cover all chokepoints with heavy firepower, and to be able to reinforce threatened sectors."[12] Indeed, the inability of the IDF's northern and southern columns to reach Karameh is directly attributable to the astute deployment of Arab Legion forces ahead of the battle. On the other hand, the Arab Legion appears not to have included PLO forces in its operational plan, an oversight that probably benefitted the IDF. And the PLO itself appears not to have prepared in a methodical fashion for the defense of Karameh, which certainly benefitted the IDF.

## *Intelligence and Reconnaissance*

Prior to the battle, the IDF assiduously collected information on Arab Legion and PLO deployments in the Jordan Valley through photographic intelligence (PHOTINT), signals intelligence (SIGINT), and human intelligence (HUMINT). The IAF conducted a number of aerial reconnaissance sorties over Jordanian and PLO positions in the days before the battle, while IDF forces placed these positions under increased surveillance by enhancing their patrol activities in the area.[13] On the eve of battle, therefore, the IDF had a firm grasp of the strength and disposition of Arab Legion and PLO defenses in the vicinity of Karameh.

Like the IDF, the Arab Legion assiduously collected information on its adversary's strength and disposition in the days before the battle. From observation posts situated on the high ground to the east of the Jordan Valley, Arab Legion intelligence officers were able to observe and identify IDF forces massing at several points along the Jordan River.[14] They were also able to observe visits to the area by senior IDF officers. The United States' Central Intelligence Agency (CIA) may also have alerted the Jordanians to the impending IDF operation, as the king of Jordan was on its payroll.[15] In any case, the IDF itself forfeited the element of surprise by dropping leaflets ahead of the battle urging the civilian population in the area to stay clear of the fighting, thereby giving away both the time and place of the operation.[16] This humanitarian gesture may well have hindered IDF forces in achieving their tasks.

## *Exploitation of Terrain*

The terrain around Karameh clearly favored the defenders. The Arab Legion made the most of its control of the high ground to the east of the Jordan Valley. It stationed its artillery forces in heavily bunkered forward positions there, not only endowing them with unobstructed fields of fire in support of armored and infantry forces in the valley below, but also shielding them effectively from IDF counterbattery fire and IAF air strikes.[17] The PLO, to the contrary, did not make good use of the terrain. While its base at Karameh had fortifications, these were apparently poorly sited and built, and therefore harmed rather than helped PLO forces in their defense of the area.[18]

The IDF, for the most part, performed reasonably well in overcoming the terrain disadvantages faced by its forces. A heliborne parachute infantry force first seized the high ground to the east of

Karameh, and then successfully attacked PLO forces from above.[19] Meanwhile, IDF armored–infantry columns took advantage of existing Jordan River bridges in order to cross a potentially troublesome water obstacle into Jordanian territory.[20] Furthermore, IDF engineering forces simultaneously constructed a pontoon bridge across the river in order to aid the flow of IDF armored and infantry forces to and from Jordanian territory.

### *Quantity and Quality of Forces*

Quantitatively speaking, Arab Legion–PLO forces, on the one hand, and IDF forces, on the other hand, were roughly equivalent during the Battle of Karameh. Arab Legion–PLO forces had more men in the Jordan Valley than IDF forces, but the former's men were spread out over a larger area than the latter's men. Likewise, the adversaries were roughly equivalent in respect of matériel. Arab Legion and IDF forces deployed similar numbers of tanks and other armored vehicles, while the former's advantage in artillery batteries was offset by the latter's recourse to airpower. Overall, then, numbers did not decide the battle.

Qualitative differences between the adversaries had more of an impact on the outcome of the IDF operation. These differences lay not in the quality of the matériel in the hands of the adversaries, but rather in the quality of the manpower of the adversaries. Specifically, highly trained and highly experienced IDF forces soundly drubbed poorly trained and very inexperienced PLO forces, despite the claim that the latter fought hard.[21] Better-trained and more experienced Arab Legion forces, to the contrary, proved to be a considerably tougher opponent for IDF forces, especially when the former fought from carefully prepared defensive positions.[22] On the one occasion in which Arab Legion forces attempted to counterattack IDF forces, however, they suffered a bad beating.

### *Command and Control*

IDF forces had a marked advantage over their adversaries in respect of command and control ($C^2$), especially when compared to PLO forces. Though a few IDF officers may have faltered during the fighting, the great majority ably led their men in battle. Many PLO officers, in contrast, simply abandoned their positions as soon as IDF forces arrived in their vicinity, leaving their men behind leaderless.[23] The conduct of Arab Legion officers fell somewhere between these two points on the $C^2$ spectrum. Jordanian officers performed competently on the defensive, but less than competently on the offensive. The efficiency of the Israeli $C^2$ system, in short, contributed to the IDF's triumph over PLO forces, as well as to its defeat of the Arab Legion counterattack, but it turned out to be less of an asset in ousting Arab Legion forces from their defensive positions.

### *Combined Arms Warfare*

The IDF practiced combined arms warfare (CAW) throughout the Battle of Karameh. Engineering forces assisted armored–infantry columns in crossing the Jordan River. IAF helicopters landed parachute infantry to the rear of the PLO base at Karameh. This parachute infantry force, as well as the armored–infantry column with which it linked up, then proceeded to cooperate in the destruction of that base. Artillery and airpower supported armor and infantry to good effect at Karameh itself, but proved to be less helpful in attempts to oust Arab Legion forces from their heavily bunkered positions on the high ground to the east of the Jordan Valley. All in all, the IDF's reasonably efficient practice of CAW during the battle contributed to its success in the hostilities.

The Arab Legion, too, practiced CAW throughout the battle. Its artillery batteries skillfully supported its armored and infantry forces in halting two of the three IDF armored–infantry columns before the latter could reach Karameh. Jordanian forces, not surprisingly, were better at CAW on the defensive than on the offensive. For their part, PLO forces seem to have been essentially ignorant of the practice of CAW. These forces apparently made no concerted attempt to integrate other arms—such as mortars, mines, and improvised explosive devices (IEDs)—beyond machine guns and assault rifles into the defense of their base at Karameh.

## Ramifications of Operation Hell

Though Operation Hell by no means resulted in the cessation of infiltration into Israel, Judea, and Samaria from Jordanian territory by PLO forces—it would take the expulsion of those forces from Jordan by the Arab Legion during the 1970 Jordanian civil war to bring an end to the phenomenon—it certainly hindered the ability of PLO insurgents/terrorists to carry out border incursions going forward. Not only had hundreds of these insurgents/terrorists been killed, wounded, or captured and their infrastructure destroyed by IDF forces during the

Battle of Karameh, but also the remaining insurgents/terrorists had had to abandon the Jordan Valley for safer areas further inland. The PLO's loss of its "sanctuary" in the valley—in tandem with the general disruption of its activities—provided the IDF with a valuable breathing space during which to upgrade its defenses along the Israeli–Jordanian frontier, which in turn rendered future infiltration attempts considerably more hazardous and costly for PLO forces.[24]

## Prelude to Operation Spring of Youth

Partly in consequence of the disappointing "payoff" yielded by its border infiltration campaign against Israel, the PLO turned its attention increasingly to international terrorism in order to promote its cause, attacking "soft" Jewish/Israeli targets—such as airliners, airports, embassies, and events—throughout the world. The slaughter of Israeli athletes at the 1972 Munich Olympic Games constituted perhaps the most notorious of these terrorist operations.[25]

In response to this attack, the Israeli government authorized a targeted-killing campaign against the leadership of Black September, the faction of the PLO that had actually been behind the slaughter. Much of what would be an effective (albeit not mistake free) campaign was carried out in various European countries by Israel's foreign intelligence agency, the MOSSAD; however, several high-ranking members of Black September were known to be hunkered down in Beirut, the capital of Lebanon. To eliminate them, the Israeli government called upon the IDF.

## Operation Spring of Youth

In the spring of 1973, the government authorized an ambitious IDF special operation (SPEC OP) whose objective would be not only the elimination of high-level Black September operatives in Beirut, but also the destruction of various PLO targets in and around the city, which the organization had chosen for its headquarters (HQ) after its expulsion from Jordan.[26] Indeed, Spring of Youth would be the biggest and most complicated SPEC OP in Israeli history up to that point in time. IDF special operations forces (SOF) and the Israel Navy (IN) had leading roles, while the IAF and MOSSAD had supporting roles.

Spring of Youth had five distinct components. First, members of Sayeret Matkal, the IDF General Staff's reconnaissance unit, would liquidate three Black September leaders. Second, members of Sayeret Tzanchanim would knock out a command post (CP) belonging to the Popular Front for the Liberation of Palestine (PFLP), another faction of the PLO. Third, members of Shayetet 13, the IN's SOF, would destroy a dual-purpose CP/arms factory belonging to Fatah, yet another faction of the PLO. Fourth, other members of Shayetet 13 would demolish a second ammunition dump belonging to Fatah. And, fifth, other members of Sayeret Tzanchanim would destroy a third ammunition dump belonging to Fatah. All of the SOF teams would be transported close to their respective targets aboard IN fast patrol boats (FPBs), and would then be ferried ashore by Zodiac boats crewed by members of Shayetet 13. The SOF teams headed to targets in downtown Beirut would be driven to their destinations from the beach by MOSSAD operatives who had rented cars specifically for this purpose. IAF helicopters would be on standby to remove any IDF dead and wounded from the field. Each of the five teams would withdraw from the target in the same manner that it had arrived there.

Carried out during the night of April 10, to make a long story short, all five raids of Operation Spring of Youth went off like clockwork. The three Black September leaders, along with dozens of PLO insurgents/terrorists, died during the operation. Black September itself would never recover from its wounds, essentially ceasing to exist as an organization thereafter. The PFLP and Fatah CPs, the Fatah arms factory, and the two Fatah ammunition dumps went up in flames as well. The IDF, on the other hand, lost just two SOF soldiers killed and three more wounded in the operation. The IAF evacuated these casualties by air, while the SOF teams were recovered and returned to Israel uneventfully by IN FPBs.

## Analysis of Operation Spring of Youth

An analysis of the operation in light of the six variables follows.

### *Operational Planning*

The IDF put together a meticulously detailed operational plan for Spring of Youth based on the elements of surprise and speed. Hence, the IDF decided to launch a night attack from the sea. And, hence, it decided to carry out all five raids simultaneously.[27] Airborne insertions would give away the element of surprise.[28] The route to and from the target of each SOF team was precisely chosen to get that team to and from its destination via the quickest and least dangerous avenue. The men of each SOF team, as well as their arms and equipment, were also selected with the utmost attention to their suitability for the operation. Furthermore, each team practiced its assignment repeatedly before the operation, developing contingency plans to address potential problems that might crop up during the operation itself.[29]

For its part, the PLO seems not to have anticipated—let alone prepared for—the possibility of an IDF operation against its assets in the Beirut area. Its defensive measures against such an operation appear to have been limited to posting sentries in front of buildings occupied by its insurgents/terrorists.

### Intelligence and Reconnaissance

The IDF's intelligence and reconnaissance effort proved to be as diligent as its planning effort. MOSSAD operatives scouted the targets in advance of the operation, noting the routines of the insurgents/terrorists stationed at those targets. They also scouted the proposed landing sites for the SOF teams. They even acquired the architectural plans of the apartments in which the Black September leaders had taken up residence.[30] The IDF undoubtedly added to the information provided by the MOSSAD via its own SIGINT and PHOTINT assets. All in all, the IDF knew what it needed to know for the operation to be a success.

The PLO, on the other hand, had no means with which to collect intelligence about or conduct surveillance of the IDF itself, let alone the IDF's preparations for this particular SPEC OP.

### Exploitation of Terrain

The IDF made effective use of the urban terrain in and around Beirut. MOSSAD operatives found stretches of beach that not only offered good cover for landing SOF teams, but also were close to their targets. The routes chosen to and from targets in downtown Beirut not only minimized the time needed to reach those targets, but also skirted as much as possible locations where Lebanese army, Lebanese police, or PLO forces were known to be stationed. The Sayeret Matkal and Sayeret Tzanchanim teams that attacked the Black September leadership and the PFLP HQ, respectively, were familiar with the layouts of the buildings occupied by their adversaries, using this knowledge to maximize the elements of surprise and speed during their assaults.

PLO forces, in contrast, did not make effective use of the urban terrain. Except for the presence of small numbers of sentries, their installations were not fortified against attack. The PLO had not constructed any sort of forward defenses to stop attacking forces before they could reach the installations themselves. Moreover, PFLP insurgents/terrorists, during their attempt to defend their HQ, unwisely used the building's elevators to get down to ground level to confront the Sayeret Tzanchanim team, which allowed the latter to mow them down in bunches before blowing up the building itself.[31]

### Quantity and Quality of Forces

PLO forces held a numerical advantage in respect of numbers of men on scene. Any they also had a distinct advantage in terms of the firepower available to them. The compact IDF SOF teams had only small arms, rifle grenades, and explosives at their disposal, while PLO forces could call on heavy machine guns and antitank weapons, as well as armored vehicles.[32] Nevertheless, this imbalance in the PLO's favor had no effect on the outcome of Spring of Youth. The much higher quality of IDF SOF soldiers vis-à-vis their insurgent/terrorist adversaries, to the contrary, had a substantial impact on the outcome of the operation.

### Command and Control

The IDF put in place a very flexible $C^2$ system during Spring of Youth. Though a central CP, with a senior officer at the helm, was set up on one of the IN's FPBs, decision-making power throughout the operation lay with the junior officers on the ground.[33] The thinking here, which undoubtedly contributed to the success of Spring of Youth, was that junior officers on the ground were in better positions than senior officers in the rear to contend with any problems that cropped up during the operation. Indeed, on at least two occasions, judicious actions by these officers worked strongly in favor of IDF forces.

PLO forces put up little or no resistance at four of the five targets attacked by IDF forces. Only the forces at PFLP HQ put up stiff resistance. Even here, however, that resistance turned out to be quite disorganized and largely ineffective. The lack of organized resistance strongly suggests that PLO forces had no dedicated $C^2$ system in place during Spring of Youth.

### Combined Arms Warfare

The IDF's sound practice of CAW at both the tactical and operational levels of warfare helped to ensure that Spring of Youth ended in a victory for Israel. Each of the five SOF teams seamlessly integrated machine guns, small arms, rifle grenades, and explosives in the context of standard IDF fire and movement protocols at the tactical level in order to defeat PLO forces. And, despite an occasional hiccup, cooperation among the SOF teams, the IN, the MOSSAD, and the IAF proved to be quite good at the operational level.

The PLO, in contrast, had no grasp of the principles that lie behind CAW. In those places in which its forces fought back, they did not integrate different arms in their defense in an effective manner.

## Ramifications of Operation Spring of Youth

Apart from inflicting a coup de grâce on Black September itself, Spring of Youth had an unmistakably negative—albeit rather temporary—impact on the PLO's ability to engage in an insurgency/terrorist campaign against Israel. The PLO's leadership appeared to be stunned by the Israeli raid. Alas, insurgent/terrorist organizations are often able to recover in reasonably short order in the wake of major defeats, and the PLO did so in the wake of Spring of Youth. Indeed, in the years after the 1973 Yom Kippur War, PLO forces in Lebanon were able to mount a string of very destructive terrorist attacks against Israel from that quarter.[34]

# CONCLUSION
# Causes of Victory, Defeat, or Draw

The following tables offer a concise recap of the eight battles profiled in this monograph.

### Summary of Battles

1 Battle of Latrun
2 First Battle of Abu Ageila/Um Katef
3 Second Battle of Abu Ageila /Um Katef
4 Operation Blossom
5 Battle of October 14
6 Operation Mole Cricket 19
7 Operation Hell
8 Operation Spring of Youth

| Battle |  |  |  | IVs |  |  | DV |
|---|---|---|---|---|---|---|---|
| Number | OP | I/R | E/T | Q/Q | C$^2$ | CAW | OIDF |
| 1 | ▯ | ≈ | ▯ | ▯ | ▯ | ▯ | Defeat |
| 2 | ≈ | ≈ | ≈ | ≈ | ≈ | ≈ | Draw |
| 3 | ◊ | ◊ | ◊ | ◊ | ◊ | ◊ | Victory |
| 4 | ▯ | ▯ | ◊ | ≈ | ◊ | ≈ | Defeat |
| 5 | ◊ | ◊ | ◊ | ◊ | ◊ | ◊ | Victory |
| 6 | ◊ | ◊ | ◊ | ◊ | ◊ | ◊ | Victory |
| 7 | ≈ | ≈ | ≈ | ◊ | ◊ | ◊ | Victory |
| 8 | ◊ | ◊ | ◊ | ◊ | ◊ | ◊ | Victory |

Key: Independent Variables (IVs): operational planning (OP); intelligence and reconnaissance (I/R); exploitation of terrain (E/T); quantity and quality of forces (Q/Q); command and control (C$^2$); and combined arms warfare (CAW).

Dependent Variable (DV): outcome for the IDF (OIDF).

◊ Indicates IDF more effective than adversary.

▯ Indicates IDF less effective than adversary.

≈ Indicates IDF and adversary equally effective.

A number of specific observations are plainly evident based on the information assembled in this table. First, when the Israel Defense Forces (IDF) scored better than its adversaries in respect of each of the six independent variables—operational planning (OP), intelligence and reconnaissance (I/R), exploitation of terrain (E/T), quantity and quality of forces (Q/Q), command and control ($C^2$), and combined arms warfare (CAW)—it obtained unequivocal victories on the battlefield. The Second Battle of Abu Ageila/Um Katef during the 1967 Six-Day War, the Battle of October 14 during the 1973 Yom Kippur War, Operation Mole Cricket 19 during the 1982 Lebanon War, and Operation Spring of Youth all resulted in such victories. Second, when the IDF scored worse than its adversary in respect of most of the independent variables, it suffered an equally unequivocal defeat on the battlefield. The Battle of Latrun in the 1947–49 War of Independence certainly ranks among the worst defeats in IDF history. Third, when the IDF and its adversary scored roughly equal marks in respect of each of the independent variables, the battle ended in a draw. The First Battle of Abu Ageila/Um Katef during the 1956 Sinai Campaign saw neither the IDF nor the Egyptian army walk away from the battlefield as either victor or vanquished.

Observations with regard to the other two battles—Operations Hell and Blossom—are less clear-cut. The former battle resulted in an Israeli victory, but not the unequivocal sort of triumph that the IDF achieved in the Second Battle of Abu Ageila/Um Katef, the Battle of October 14, Mole Cricket 19, and Spring of Youth. The less-than-decisive nature of the Israeli victory can be chalked up to the skill displayed by the Arab Legion with respect to OP, I/R, and E/T. The explanation behind the latter battle is a bit more complex. Though the IDF triumphed at the tactical level—the Israel Air Force (IAF) deep-penetration raids during the 1969–70 War of Attrition succeeded in destroying their targets—the battle itself nevertheless resulted in an Israeli defeat at the operational level, because it triggered a massive Soviet intervention in the fighting that eventually led to a major setback for Israel in respect of the outcome of the war. The main culprits behind the Israeli defeat at the operational level lie in poor OP and I/R, both of which completely misjudged the Egyptian, Soviet, and American responses to the IAF's deep-penetration bombing campaign.

## General Observations

Based on the totality of the evidence furnished by this sample of eight battles, the "model" set forth in this monograph would appear to contain a considerable amount of explanatory power. Collectively speaking, it seems that the six independent variables incorporated into this model capture much of the explanation behind the causes of victory, defeat, or draw on the battlefield. What this model cannot determine with any precision are the relationships of these variables to one another. The outcomes of the two battles in which the independent variables were mixed in the sense that they did not clearly favor one adversary over the other one strongly suggest that the relative importance of particular variables in determining the outcomes of particular battles vary from battle to battle. To determine which particular variables are most important with respect to any particular battle requires a far more rigorous and sophisticated model than the one proffered here. Finally, though this model has been applied only to the case of the IDF from 1948 to 1982, there are no prima facie reasons to suspect that it does not apply equally well to the cases of other national armies. Nor are there any prima facie reasons to suspect that the model does not apply equally well to a world in which asymmetrical wars have largely supplanted interstate wars.

# Glossary of Technical Terms

**Airborne Electronic Warfare (AEW) Platform**: an aircraft specially equipped to engage in electronic warfare (EW) tasks, such as jamming radar and/or communications systems, in order to impede an adversary's ability to conduct air, naval, and/or ground operations.

**Airborne Warning and Control System (AWACS) Platform**: an aircraft specially equipped to serve as a command and control ($C^2$) post in the sky, monitoring the location and disposition of an adversary's air, naval, and/or ground forces, as well as directing friendly air, naval, and/or ground forces to their targets.

**Air Superiority**: the attainment of control of the airspace above the battlefield.

**Antiaircraft Artillery (AAA)**: artillery that is optimized to shoot down aircraft. It may be guided by radar or optical systems, and it may be deployed in single units or in batteries made up of multiple units.

**Antiradiation Missile (ARM)**: a missile designed exclusively to destroy an adversary's radar units.

**Arm**: a specific combat branch of the armed forces, such as armor, artillery, infantry, engineers, intelligence, air force, navy, and so on. The term is sometimes also used to refer to noncombat branches, such as logistics and personnel.

**Armored Fighting Vehicle (AFV)**: any armored vehicle that has a role on the battlefield.

**Armored Personnel Carrier (APC)**: an armored vehicle designed for use by the infantry as a fighting platform on the battlefield; however, it may also be used simply to transport infantry to or around the battlefield rather than as a fighting platform.

**Asymmetrical War (AW)**: a sustained high-intensity conflict between (or among) a state (or states), on the one hand, and an insurgent/terrorist group (or groups), on the other hand.

**Attrition Warfare**: a mode of warfare that favors firepower over mobility.

**Close Air Support (CAS)**: air strikes mounted against adversarial ground forces in close proximity to friendly ground forces in order to support the latter.

**Cluster Bomb Unit (CBU)**: a bomb that disperses a large number of bomblets over a wide area. Such bombs are often employed against personnel, but they may also be used against tanks, APCs, artillery, and the like.

**Combat Air Patrol (CAP)**: a patrol by friendly aircraft intended to keep hostile aircraft out of a particular airspace.

**Combined Arms Warfare (CAW)**: a mode of warfare that is based on the premise that better results on the battlefield will be achieved by relying on a mix of different arms rather than on a single arm. Both balanced (symmetrical) and unbalanced (asymmetrical) versions of CAW have been tested on the battlefield.

**Command and Control ($C^2$)**: the principles and process behind the management of friendly ground, air, and/or naval forces in order to achieve a specified objective (or set of objectives) on the battlefield.

**Command, Control, Communications, and Intelligence ($C^3I$)**: a more sophisticated version of $C^2$.

**Defensive Warfare**: responding to the initiative of an adversary on the battlefield.

**Electronic Warfare (EW)**: the employment of electronic systems by friendly forces to disrupt the ability of adversarial forces to conduct operations on the battlefield.

**Fast Patrol Boat (FPB)**: a type of small, fast, and maneuverable warship often fitted with antiship missiles in order to allow it to engage similar or larger naval vessels.

**Ground Control Intercept (GCI)**: a $C^2$ system in which air controllers on the ground vector friendly aircraft to intercept hostile aircraft.

**Human Intelligence (HUMINT)**: information about an adversary collected by human agents.

**Improvised Explosive Device (IED)**: a homemade bomb cobbled together from various spare parts and explosive substances.

**Integrated Air Defense System (IADS)**: an air defense system that combines SAM and AAA batteries, as well as their associated search and fire-control radars, into a seamless web in an effort to deny an opposing air force the ability to attack the area the IADS is intended to defend.

**Interstate War**: a sustained high-intensity conflict between (or among) states.

**Low-intensity conflict (LIC)**: persistent low-level hostilities between (or among) a state (or states), on the one hand, and a state (or states), an insurgent/terrorist group (or groups), a hostile population (or populations), or some combination of these entities, on the other hand.

**Maneuver Warfare**: a mode of warfare that favors mobility over firepower.

**Mechanized Infantry**: infantry that is mounted on and fights from APCs. For the sake of terminological simplicity, this term also covers armored infantry, a special category of mechanized infantry that trains specifically to support an army's tank forces and is directly incorporated into those forces.

**Nap of the Earth (NOE)**: a low-level aircraft flight path that uses the terrain to mask the aircraft from observation by the adversary.

**Noncommissioned officer (NCO)**: an officer who obtains his/her position as a result of advancement through the enlisted ranks rather than as a result of receiving a commission upon graduation from officer training. In practice, noncommissioned officers are those officers who occupy ranks below the rank of lieutenant.

**Nonmechanized Infantry**: infantry that fights primarily on foot. Such infantry is sometimes also referred to as "motorized" in those instances when it is transported to and around the battlefield in armored vehicles and/or trucks. Nevertheless, because this type of infantry typically dismounts before going into action, nonmechanized infantry is a more suitable term for it.

**Offensive Warfare**: taking the initiative against an adversary on the battlefield.

**Operational Level of Warfare**: the principles according to which larger military formations (i.e., divisions and armies) are organized for and employed in battle.

**Order of Battle (ORBAT)**: the entire roster of equipment and units possessed by an army, or a circumscribed roster of equipment and units deployed by it in a particular battle.

**Photographic Intelligence (PHOTINT)**: information about an adversary collected by photographic means.

**Precision-guided Munition (PGM)**: a bomb, missile, or shell that is guided to a target by one (or more) of several means, including radar, television camera, laser beam, and/or ground positioning system (GPS) coordinates.

**Revolution in Military Affairs (RMA)**: a way of war that emphasizes the advantages to be derived from the application of advanced technologies on the battlefield.

**Signals Intelligence (SIGINT)**: information about an adversary collected by electronic means, such as intercepting radio communications or tapping telephone wires.

**Special Operation (SPEC OP)**: a sensitive, high-risk operation, often carried out deep inside an adversary's territory, by specially trained and equipped forces.

**Special Operations Force (SOF)**: a force trained and equipped to carry out sensitive, high-risk operations, often deep inside an adversary's territory.

**Suppression of Enemy Air Defenses (SEAD)**: a mission assigned to an air force, often in collaboration with ground forces, whose objective is the destruction of an adversary's air defenses.

**Surface-to-air Missile (SAM)**: a missile that is optimized to shoot down aircraft. It is typically deployed as part of a battery that consists of several missile launchers and a fire-control radar unit. A single soldier can operate small, shoulder-launched missiles, which are typically heat seeking.

**Surface-to-surface Missile (SSM)**: a missile that is optimized to destroy ground targets. This category of weapon includes short-range missiles, medium-range missiles, and long-range missiles.

**Tactical Level of Warfare**: the principles according to which smaller military formations (i.e., squads through brigades) are organized for and employed in battle.

**Unmanned Aerial Vehicle (UAV)**: a drone that is flown remotely by a pilot from a ground station. Historically, UAVs have been deployed primarily in order to gather intelligence about an adversary; however, over the past couple of decades, they have been employed increasingly to launch attacks against an adversary.

**Weapons of Mass Destruction (WMD)**: weapons that are capable of inflicting a massive number of casualties with a single use. Chemical, biological, and nuclear weapons fall into this category, while conventional weapons are traditionally exempted from it.

# Notes

## Introduction

1 Representative of this school of thought is Victor Davis Hanson, *Carnage and Culture: Landmark Battles in the Rise of Western Power* (New York: Doubleday, 2001). In the context of the Arab–Israeli conflict, see Kenneth M. Pollack, *Armies of Sand: The Past, Present, and Future of Arab Military Effectiveness* (New York: Oxford University Press, 2019).

2 Representative of this school of thought is Stephen Biddle, *Military Power: Explaining Victory and Defeat in Modern Battle* (Princeton, NJ: Princeton University Press, 2004).

3 Representative of this school of thought is John Ellis, *Brute Force: Allied Strategy and Tactics in the Second World War* (New York: Viking, 1990).

4 Representative of this school of thought is Harold A. Winters et al., *Battling the Elements: Weather and Terrain in the Conduct of War* (Baltimore: Johns Hopkins University Press, 1998).

5 Though the Arab–Israeli conflict has witnessed several naval battles during its history, especially during the Yom Kippur War, naval warfare has never been more than a marginal phenomenon in Arab–Israeli hostilities, which is why naval warfare is not probed in this monograph.

## 1 The Battle of Latrun

1 A compact account of the Yishuv's experience with self-defense prior to the establishment of Israel can be found in Ze'ev Schiff, *A History of the Israeli Army: 1874 to the Present*, Rev. ed. (New York: Macmillan Publishing Company, 1985), pp. 1–20.

2 For a comprehensive history of the Haganah, see Ben-Zion Dinur (ed.), *History of the Hagana*, Vols. 1–3 (Tel Aviv: Am Oved Publishers, 1973).

3 An authoritative review of the policies and actions of the Jewish militias during the decades of British rule in Palestine, with an emphasis on the policies and actions of the dissident organizations, is delivered in Bruce Hoffman, *Anonymous Soldiers: The Struggle for Israel, 1917–1947* (New York: Knopf, 2015).

4 For British–Jewish cooperation during the former, see Matthew Hughes, *Britain's Pacification of Palestine: The British Army, the Colonial State, and the Arab Revolt, 1936–1939* (Cambridge: Cambridge University Press, 2019). For British–Jewish cooperation during the latter, see Dinur, *History of the Hagana*.

5 On armed Jewish resistance to the British, see Hoffman, *Anonymous Soldiers*, and Saul Zadka, *Blood in Zion: How the Jewish Guerrillas Drove the British Out of Palestine* (London: Brassey's, 1995).
6 The history of the Jewish Brigade is recounted in Morris Beckman, *The Jewish Brigade: An Army with Two Masters, 1944–45* (New York: De Capo Press, 1998).
7 Authoritative accounts of the War of Independence can be found in Chaim Herzog, *The Arab–Israeli Wars: War and Peace in the Middle East from the War of Independence Through Lebanon*, Rev. ed. (New York: Vintage Books, 1984), pp. 17–104, and Benny Morris, *1948: A History of the First Arab–Israeli War* (New Haven: Yale University Press, 2008).
8 For Operation Nachshon and Plan D, see Herzog, *The Arab–Israeli Wars*, pp. 29–38, and Morris, *1948*, pp. 113–179.
9 On the IDF ORBAT at the beginning of the interstate phase of the war, see Herzog, *The Arab–Israeli Wars*, p. 48, and Kenneth M. Pollack, *Arabs at War: Military Effectiveness, 1948–2001* (Lincoln, NE: University of Nebraska Press, 2002), pp. 15–16.
10 On the combined Arab forces ORBAT at the beginning of the interstate phase of the war, see Herzog, *The Arab–Israeli Wars*, pp. 47–48, and Pollack, *Arabs at War*, pp. 15–16, 150, 270–271, and 448.
11 The fighting at Latrun is traced in Herzog, *The Arab–Israeli Wars*, pp. 63–68 and 83; Morris, *1948*, pp. 219–231; and Pollack, *Arabs at War*, pp. 276–282. The present account of the battle is compiled from these sources.
12 Morris, *1948*, p. 220, and Pollack, *Arabs at War*, p. 277.
13 For the Arab Legion ORBAT at Latrun, see Herzog, *The Arab–Israeli Wars*, p. 64; Morris, *1948*, p. 221; and Pollack, *Arabs at War*, pp. 276–277.
14 For the IDF ORBAT at Latrun, see Morris, *1948*, p. 221, and Pollack, *Arabs at War*, p. 277.
15 Accounts of the IDF's first attempt to capture Latrun are related in Herzog, *The Arab–Israeli Wars*, pp. 64–65; Morris, *1948*, pp. 221–224; and Pollack, *Arabs at War*, p. 277.
16 IDF and Arab Legion casualties are recorded in Morris, *1948*, p. 224.
17 Accounts of the IDF's second attempt to capture Latrun are related in Herzog, *The Arab–Israeli Wars*, pp. 65–66; Morris, *1948*, pp. 224–229; and Pollack, *Arabs at War*, pp. 277–278.
18 Herzog, *The Arab–Israeli Wars*, p. 65; Morris, *1948*, pp. 225–226; and Pollack, *Arabs at War*, pp. 277–278.
19 Herzog, *The Arab–Israeli Wars*, pp. 65–66; Morris, *1948*, pp. 227–229; and Pollack, *Arabs at War*, p. 278.
20 IDF and Arab Legion casualties are recorded in Morris, *1948*, p. 229.
21 On this road, see Herzog, *The Arab–Israeli Wars*, pp. 66–68; Morris, *1948*, pp. 230–231; and Pollack, *Arabs at War*, p. 278.

22 Accounts of the IDF's third attempt to capture Latrun are related in Herzog, *The Arab–Israeli Wars*, p. 68; Morris, *1948*, pp. 229–230; and Pollack, *Arabs at War*, p. 278.
23 Accounts of the IDF's fourth attempt to capture Latrun are related in Herzog, *The Arab–Israeli Wars*, p. 83, and Pollack, *Arabs at War*, pp. 280–282.
24 Ibid.
25 For total IDF manpower losses during the Battle of Latrun, see the battle's Wikipedia web page at <en.wikipedia.org/wiki/Battles_of_Latrun_(1948)>.
26 Morris, *1948*, p. 221.
27 Ibid.
28 Herzog, *The Arab–Israeli Wars*, p. 68; Morris, *1948*, p. 229; and Pollack, *Arabs at War*, pp. 277 and 280–281.
29 Herzog, *The Arab–Israeli Wars*, p. 63; Morris, *1948*, pp. 220–221; and Pollack, *Arabs at War*, p. 277.
30 Pollack, *Arabs at War*, p. 277.
31 Morris, *1948*, pp. 220–221.
32 Pollack, *Arabs at War*, pp. 282–284.
33 Morris, *1948*, p. 222.
34 Ibid., p. 230.
35 Ibid., pp. 223–224.
36 Pollack, *Arabs at War*, p. 282.
37 Morris, *1948*, pp. 231–232, and Pollack, *Arabs at War*, p. 282.

## 2 The First Battle of Abu Ageila / Um Katef

1 An in-depth review of the human, economic, and social costs of the war for Israel can be found in Moshe Naor, *Social Mobilization in the Arab–Israeli War of 1948: On the Israeli Home Front* (London: Routledge, 2013).
2 Eastern Palestine—that is, the approximately 80 percent of the territory originally allotted to the mandate that lay east of the Jordan River—had been transformed by British government decree into the Hashemite Kingdom of Jordan during the early 1920s.
3 The pre-Sinai Campaign debate within the Israeli defense establishment over IDF force structure is reviewed in F. H. Toase, "The Israeli Experience of Armoured Warfare," pp. 163–164, in J. P. Harris and F. H. Toase (eds.), *Armoured Warfare* (New York: St. Martin's Press, 1990), and Alaric Searle, *Armoured Warfare: A Military, Political and Global History* (London: Bloomsbury, 2017), p. 155.
4 Yagil Henkin, *The 1956 Suez War and the New World Order in the Middle East: Exodus in Reverse* (Lanham, MD: Lexington Books, 2015), pp. 136–137.
5 On hostilities between the IDF and the Egyptian army in the early 1950s, see Michael B. Oren, "Escalation to Suez: The Egypt–Israel Border War,

1949–56," *Journal of Contemporary History*, Vol. 24, No. 2 (April 1989), pp. 347–375, and Jonathan Shimshoni, *Israel and Conventional Deterrence: Border Warfare from 1953 to 1970* (Ithaca, NY: Cornell University Press, 1988), pp. 69–122.

6 On the Egyptian sea blockade, see Shimshoni, *Israel and Conventional Deterrence*, pp. 71–74. Egypt also repeatedly denied ships to or from Israel the use of the Suez Canal.

7 A thorough treatment of the countdown to the Sinai Campaign can be found in Henkin, *The 1956 Suez War*, pp. 29–118.

8 For the specifics of the arms deal, see Henkin, *The 1956 Suez War*, p. 48; Searle, *Armoured Warfare*, p.155; and Shimshoni, *Israel and Conventional Deterrence*, pp. 104–105.

9 On the debate within the Israeli defense establishment over how to respond to the arms deal, see Motti Golani, *Israel in Search of a War: The Sinai Campaign, 1955–1956* (Brighton: Sussex Academic Press, 1998).

10 For the French and British reasons, see Henkin, *The 1956 Suez War*, pp. 69–89.

11 For the Franco–British–Israeli agreement to invade Egypt, see ibid., pp. 91–107.

12 The IDF–Egyptian army balance of forces on the eve of the war is outlined in Henkin, *The 1956 Suez* War, pp. 114–115; Chaim Herzog, *The Arab–Israeli Wars: War and Peace in the Middle East from the War of Independence Through Lebanon*, Rev. ed. (New York: Vintage Books, 1984), pp. 115 and 117; and Kenneth M. Pollack, *Arabs at War: Military Effectiveness, 1948–1991* (Lincoln, NE: University of Nebraska Press), p. 31.

13 Accounts of Operation Kadesh 2 can be found in Chaim Herzog, *The Arab–Israeli* Wars, pp. 114–138; Henkin, *The 1956 Suez War*, pp. 119–127, 135–156, 159–191, 197, and 254; and Pollack, *Arabs at War*, pp. 32–40.

14 On the Israeli capture of the Abu Ageila/Um Katef complex during the War of Independence, see Benny Morris, *1948: The First Arab–Israeli War* (New Haven, Yale University Press, 2008), p. 363, and Henkin, *The 1956 Suez War*, p. 135.

15 Thorough treatments of the First Battle of Abu Ageila/Um Katef can be found in George W. Gawrych, *Key to the Sinai: The Battles for Abu Ageila in the 1956 and 1967 Arab–Israeli Wars* (Fort Leavenworth, KS: U.S. Army Command and General Staff College, 1990), pp. 1–65, and Henkin, *The 1956 Suez War*, pp. 135–156. Additional accounts of the battle can be found in Herzog, *The Arab–Israeli Wars*, pp. 123–129, and Pollack, *Arabs at War*, pp. 32–35. The present description of the battle is drawn from these sources.

16 Egyptian army and IDF ORBATs in respect of the battle can be found in Gawrych, *Key to the Sinai*, pp. 14 and 24; Henkin, *The 1956 Suez War*, p. 136; and Herzog, *The Arab–Israeli Wars*, pp. 124 and 127.

17 Gawrych, *Key to the Sinai*, pp. 36–41, and Henkin, *The 1956 Suez War*, p. 138.
18 Gawrych, *Key to the Sinai*, pp. 41–44, and Herzog, *The Arab–Israeli Wars*, pp. 125–126.
19 Gawrych, *Key to the Sinai*, pp. 44–50, and Henkin, *The 1956 Suez War*, pp. 139–144.
20 Gawrych, *Key to the Sinai*, pp. 50–52 and 57–58; Henkin, *The 1956 Suez War*, pp. 147–153; and Herzog, *The Arab–Israeli Wars*, pp. 128–129.
21 Gawrych, *Key to the Sinai*, pp. 54 and 57; Henkin, *The 1956 Suez War*, pp. 145–147; Herzog, *The Arab–Israeli Wars*, pp. 127–128; and Pollack, *Arabs at War*, pp. 34–35.
22 Gawrych, *Key to the Sinai*, pp. 59–60, and Henkin, *The 1956 Suez War*, pp. 153–155.
23 Gawrych, *Key to the Sinai*, p. 64, and Henkin, *The 1956 Suez War*, p. 154.
24 Gawrych, *Key to the Sinai*, p. 26.
25 In case France and Great Britain were to get cold feet at the last minute and refrain from initiating hostilities against Egypt, the Israeli government wanted to be able to present its opening military moves in the Sinai as a limited raid rather than a full-fledged invasion, which would provide it with a face-saving option to withdraw IDF forces from the peninsula. The early entry of IDF armor into battle would negate this option, as such a move would be interpreted by Egypt and the international community as an unequivocal commitment to full-scale war. Henkin, *The 1956 Suez War*, pp. 137–138, and Herzog, *The Arab–Israeli Wars*, p. 117.
26 Gawrych, *Key to the Sinai*, p. 30.
27 Gawrych, *Key to the Sinai*, p. 24; Henkin, *The 1956 Suez War*, p. 148; and Hezog, *The Arab–Israeli Wars*, p. 128.
28 Gawrych, *Key to the Sinai*, pp. 33–36; Herzog, *The Arab–Israeli Wars*, p. 118; and Pollack, *Arabs at War*, pp. 41–42.
29 Gawrych, *Key to the Sinai*, pp. 14–15 and 17–19.
30 Ibid., p. 30.
31 Henkin, *The 1956 Suez War*, pp. 145 and 149–150, and Herzog, *The Arab–Israeli Wars*, p. 128.
32 Gawrych, *Key to the Sinai*, p. 41; Henkin, *The 1956 Suez War*, pp. 140–141; and Herzog, *The Arab–Israeli Wars*, p. 126.
33 Descriptions of the natural terrain in and around Abu Ageila/Um Katef can be found in Gawrych, *Key to the Sinai*, pp. 9 and 13–14, and Eric Hammel, *Six Days in June: How Israel Won the 1967 Arab–Israeli War* (New York: Scribner's, 1992), pp. 229–231.
34 Descriptions of the 6th (EG) Infantry Brigade's defenses at Abu Ageila/Um Katef can be found in Gawrych, *Key to the Sinai*, pp. 14–19, and Herzog, *The Arab–Israeli Wars*, pp. 124 and 127.
35 The opposing numbers of men and machines are highlighted in Gawrych, *Key to the Sinai*, p. 25.

36 For a list of the arms possessed by each side, see Gawrych, *Key to the Sinai*, pp. 17 and 25.
37 Henkin, *The 1956 Suez War*, p. 144; Herzog, *The Arab–Israeli Wars*, p. 127; and Pollack, *Arabs at War*, p. 37.
38 Gawrych, *Key to the Sinai*, pp. 50–52 and 57–58, and Henkin, *The 1956 Suez War*, pp. 148–151.
39 Gawrych, *Key to the Sinai*, p. 58, and Henkin, The *1956 Suez War*, pp. 151–153.
40 Gawrych, *Key to the Sinai*, p. 19, is perhaps a bit overly enthusiastic in his assessment of the quality of the 6th (EG) Infantry Brigade.
41 On IDF $C^2$ problems throughout the battle, see Gawrych, *Key to the Sinai*, pp. 36, 40, and 52–54, and Henkin, *The 1956 Suez War*, pp. 137–139 and 147–153.
42 Quoted in Gawrych, *Key to the Sinai*, p. 40, and Henkin, *The 1956 Suez War*, p. 139.
43 On this friendly fire incident, see Gawrych, *Key to the Sinai*, p. 64, and Henkin, *The 1956 Suez War*, p. 155. Faulty air–ground cooperation throughout the war also caused friendly fire incidents between the IAF and IDF ground forces.
44 Gawrych, *Key to the Sinai*, p. 45, and Henkin, *The 1956 Suez War*, p. 155.
45 On the Egyptian $C^2$ system at Abu Ageila/Um Katef, see Gawrych, *Key to the Sinai*, pp. 14 and 19.
46 Henkin, *The 1956 Suez War*, p. 153.
47 Ibid., p. 143.
48 For estimates of Israeli and Egyptian losses, see Henkin, *The 1956 Suez War*, p. 254, and Pollack, *Arabs at War*, p. 39.

## 3 The Second Battle of Abu Ageila/Um Katef

1 George W. Gawrych, *Key to the Sinai: The Battles of Abu Ageila in the 1956 and 1967 Arab–Israeli Wars* (Fort Leavenworth, KS: U.S. Army Command and General Staff College, 1990), p. 67.
2 Ibid., p. 68.
3 On the aircraft buildup from 1956–67, see Eliezer Cohen, *Israel's Best Defense: The First Full Story of the Israel Air Force* (New York: Orion Books, 1993), pp. 105 and 194, and Gawrych, *Key to the Sinai*, p. 67.
4 On the tank buildup, see F. H. Toase, "The Israeli Experience of Armoured Warfare," pp. 163 and 167, in J. P. Harris and F. H. Toase (eds.), *Armoured Warfare* (New York: St. Martin's Press, 1990), and Alaric Searle, *Armoured Warfare: A Military, Political and Global History* (London: Bloomsbury, 2017), pp. 155–157.
5 Jonathan M. House, *Combined Arms Warfare in the Twentieth Century* (Lawrence, KS: University Press of Kansas, 2001), p. 230.
6 For brief summaries of the IDF's war-fighting doctrine on the eve of the Six-Day War, see Gawrych, *Key to the Sinai*, pp. 67–69, and David

Rodman, *Combined Arms Warfare in Israeli Military History: From the War of Independence to Operation Protective Edge* (Brighton: Sussex Academic Press, 2019), pp. 20–22.

7 While spot on in its broad strokes, Gawyrch's assessment of the IDF's pre-1967 war-fighting doctrine—like many other informed assessments of the pre-1967 doctrine—actually more accurately reflects the IDF's post-1967 doctrine in regard to the role of CAW.

8 House, *Combined Arms Warfare*, p. 230.

9 Rodman, *Combined Arms Warfare*, pp. 23–29.

10 Ibid., p. 22.

11 The most authoritative account of the Six-Day War, including the train of military and diplomatic events leading up to the hostilities, is Michael B. Oren, *Six Days of War: June 1967 and the Making of the Modern Middle East* (New York: Oxford University Press, 2002). For a different interpretation of these events, see Guy Laron, *The Six-Day War: The Breaking of the Middle East* (New Haven: Yale University Press, 2017).

12 For the IDF ORBAT on the Sinai front, see Simon Dunstan, *The Six Day War 1967*, Vol. 1 (Oxford: Osprey Publishing, 2009), p. 27, and Eric Hammel, *Six Days in June: How Israel Won the 1967 Arab–Israeli War* (New York: Scribner's, 1992), pp. 428–430.

13 See Oren, *Six Days of War*, pp. 92–97, for the Egyptian army's plan to strike first.

14 For the Egyptian army ORBAT, see Dunstan, *The Six Day War*, p. 27, and Hammel, *Six Days in June*, p. 432.

15 For the numbers of troops and tanks deployed by both the IDF and the Egyptian army, see Dunstan, *The Six Day War*, p. 26; Kenneth M. Pollack, *Arabs at War: Military Effectiveness, 1948–1991* (Lincoln, NE: University of Nebraska Press, 2002), pp. 58–59; and Searle, *Armoured Warfare*, p. 157.

16 An exhaustive account of this operation can be found in Shlomo Aloni, *The June 1967 Six-Day War: Operation Focus*, Vol. A (Bat Hefer: Isradecal, 2008).

17 Estimates of the damage suffered by the Egyptian air force on June 5 can be found in Aloni, *The June 1967 Six-Day War*, pp. 179–180, and Oren, *Six Days of War*, p. 176.

18 Aloni, *The June 1967 Six-Day War*, pp. 54 and 179.

19 The northernmost break-in battle is chronicled in Dunstan, *The Six Day War*, pp. 40–55; Hammel, *Six Days in June*, pp. 172–218; Oren, *Six Days of War*, pp. 178–181; and Pollack, *Arabs at War*, pp. 64–67.

20 The centermost break-in battle is chronicled in Dunstan, *The Six Day War*, pp. 55–57 and 60–61; Hammel, *Six Days in June*, pp. 219–227; Oren, *Six Days of War*, pp. 211–212; and Pollack, *Arabs at War*, p. 71.

21 For the 38th (IS) Armored Division's ORBAT, see Dunstan, *The Six Day War*, p. 27; Gawrych, *Key to the Sinai*, p. 92; and Hammel, *Six Days in June*, pp. 429–430.

22 For the 2nd (EG) Infantry Division's ORBAT, see Gawrych, *Key to the Sinai*, p. 82.
23 For the disposition of Egyptian forces in and around Abu Ageila/Um Katef, see ibid., pp. 82–86.
24 The present account of the Second Battle of Abu Ageila/Um Katef is compiled from the following sources: Dunstan, *The Six Day War*, pp. 58–59 and 61–69; Gawrych, *Key to the Sinai*, pp. 87–122; Hammel, *Six Days in June*, pp. 228–242; and Pollack, *Arabs at War*, pp. 67–70.
25 Dunstan, *The Six Day War*, p. 58 and 66; Gawrych, *Key to the Sinai*, p. 101 and 105; Hammel, *Six Days in June*, p. 241; and Pollack, *Arabs at War*, pp. 69–70.
26 Dunstan, *The Six Day War*, p. 58; Gawrych, *Key to the Sinai*, pp. 103–104; and Hammel, *Six Days in June*, pp. 232–234.
27 Dunstan, *The Six Day War*, pp. 58 and 65–66; Gawrych, *Key to the Sinai*, pp. 101–103; and Hammel, *Six Days in June*, pp. 234–235.
28 Dunstan, *The Six Day War*, pp. 66–68, Gawrych, *Key to the Sinai*, pp. 104–105 and 107–109; and Hammel, *Six Days in June*, pp. 235–236 and 238–239.
29 Dunstan, *The Six Day War*, pp. 66–69; Gawrych, *Key to the Sinai*, pp. 109–114; and Hammel, *Six Days in June*, pp. 237–238.
30 Dunstan, *The Six Day War*, pp. 68–69; Gawrych, *Key to the Sinai*, pp. 114–117; and Hammel, *Six Days in June*, pp. 239–241.
31 For IDF losses, see Dunstan, *The Six Day War*, p. 58; Gawrych, *Key to the Sinai*, p. 117; and Hammel, *Six Days in June*, p. 241.
32 For Egyptian army losses, see Gawrych, *Key to the Sinai*, p. 117, and Hammel, *Six Days in June*, p. 241.
33 Dunstan, *The Six Day War*, p. 65; Gawrych, *Key to the Sinai*, pp. 72–73, and Pollack, *Arabs at War*, p. 67.
34 The IDF operational plan is laid out in its essentials in Dunstan, *The Six Day War*, pp. 66–67, and Gawrych, *Key to the Sinai*, pp. 91–97.
35 The Egyptian army's preparations at Abu Ageila/Um Katef before the Six-Day War are chronicled in Gawrych, *Key to the Sinai*, pp. 80–86.
36 Dunstan, *The Six Day War*, p. 65.
37 Gawrych, *Key to the Sinai*, p. 81.
38 Hammel, *Six Days in June*, pp. 233 and 235.
39 Descriptions of the natural terrain in and around Abu Ageila/Um Katef can be found in Gawrych, *Key to the Sinai*, pp. 9 and 13–14; and Hammel, *Six Days in June*, pp. 229–231.
40 For the layout of the defenses at Abu Ageila/Um Katef on the eve of the Six-Day War, see Dunstan, *The Six Day War*, pp. 61 and 65; Gawrych, *Key to the Sinai*, pp. 82–86; and Hammel, *Six Days in June*, pp. 229–231.
41 On the numbers and disposition of men and machines fielded by the IDF and the Egyptian army at Abu Ageila/Um Katef, see Gawrych, *Key to the Sinai*, pp. 82, 93, and 98, and Pollack, *Arabs at War*, p. 68.

42 The newer model Soviet tanks that equipped the bulk of Egyptian army forces during the Six-Day War, however, were more or less the equal of the British and American tanks that equipped the bulk of IDF forces during the war.

43 For a general comparison of Arab and Israeli manpower across the length and breadth of the Arab–Israeli conflict, see Pollack, *Arabs at War*, and Kenneth M. Pollack, *Armies of Sand: The Past, Present, and Future of Arab Military Effectiveness* (New York: Oxford University Press, 2019).

44 Gawrych, *Key to the Sinai*, p. 91.

45 On IDF GHQ's reservations about a night attack, see Gawrych, *Key to the Sinai*, p. 105, and Hammel, *Six Days in June*, p. 235.

46 For an argument to the effect that junior officers across all of the Arab armies have habitually been unable to act autonomously, aggressively, and creatively as a consequence of certain behavioral traits transmitted via Arab culture, see Pollack, *Armies of Sand.*

47 Gawrych, *Key to the Sinai*, pp. 97 and 99–100.

48 Pollack, *Arabs at War*, p. 69.

49 House, *Combined Arms Warfare*, p. 230.

50 Oren, *Six Days of War*, p. 214, and Pollack, *Arabs at War*, p.71.

51 The post-Red Sheet phase of the IDF offensive is chronicled in Dunstan, *The Six Day War*, pp. 71–87; Hammel, *Six Days in June*, pp. 243–279; and Pollack, *Arabs at War*, pp. 71–74.

52 Estimates of Egyptian army and IDF losses in the Six-Day War can be found in Dunstan, *The Six Day War*, p. 88; Hammel, *Six Days in June*, p. 279; Oren, *Six Days of War*, pp. 305–306; and Pollack, *Arabs at War*, p. 74.

## 4 Operation Blossom

1 For the traditional IDF attitude toward "doing" versus "thinking," see Dima Adamsky, *The Culture of Military Innovation: The Impact of Cultural Factors on the Revolution in Military Affairs in Russia, the US, and Israel* (Palo Alto: Stanford University Press, 2010); Avi Kober, *Practical Soldiers: Israel's Military Thought and Its Formative Factors* (Leiden: Brill, 2015); and Eitan Shamir, *Transforming Command: The Pursuit of Mission Command in the U.S., British, and Israeli Armies* (Palo Alto: Stanford University Press, 2011).

2 Jonathan M. House, *Combined Arms Warfare in the Twentieth Century* (Lawrence, KS: University Press of Kansas, 2001), p. 230. Incidentally, much of the international defense community readily endorsed the Israeli defense establishment assessment of the Six-Day War.

3 The Israeli government's attitude toward the return of Judea and Samaria in the aftermath of the Six-Day War is probed in significant depth in Avshalom Rubin, *The Limits of the Land: How the Struggle for the West Bank*

*Shaped the Arab–Israeli Conflict* (Bloomington, IN: Indiana University Press, 2017), especially chs. 6 and 7.

4 The Israeli government's attitude toward the return of the Sinai and the Golan in the aftermath of the Six-Day War is noted in Aaron S. Klieman, *Israel and the World After 40 Years* (Washington, D.C.: Pergamon-Brassey's International Defense Publishers, 1990), p. 75.

5 See ibid. on the summit as it pertained to Israel.

6 Abba Eban, *Personal Witness: Israel Through My Eyes* (New York: G.P. Putnam's Sons, 1984), p. 484.

7 The first recorded instance of Egyptian fire across the canal occurred on July 1, 1967. See Chaim Herzog, *The Arab–Israeli Wars: War and Peace in the Middle East from the War of Independence Through Lebanon*, Rev. ed. (New York: Vintage Books, 1984), p. 196.

8 Yaacov Bar-Siman-Tov, *The Israeli–Egyptian War of Attrition, 1969–1970: A Case Study of Limited Local War* (New York: Columbia University Press, 1980), p. 45, and Jonathan Shimshoni, *Israel and Conventional Deterrence: Border Warfare from 1953 to 1970* (Ithaca, NY: Cornell University Press, 1988), pp. 138–141.

9 Amos Amir, *Fire in the Sky: Flying in Defence of Israel* (South Yorkshire: Pen and Sword Books, 2005), pp. 46–56.

10 The stages of the war are chronicled in Bar-Siman-Tov, *The Israeli–Egyptian War of Attrition*, pp. 43–185; Herzog, *The Arab–Israeli Wars*, pp. 195–202 and 207–219; and Shimshoni, *Israel and Conventional Deterrence*, pp. 123–211.

11 Bar-Siman-Tov, *The Israeli–Egyptian War of Attrition*, pp. 69–70.

12 For a description of the Bar Lev Line, see Simon Dunstan, *Israeli Fortifications of the October War 1973* (Oxford: Osprey Publishing, 2008), pp. 16–19.

13 Bar-Siman-Tov, *The Israeli–Egyptian War of Attrition*, p. 87.

14 Shlomo Aloni, *Israeli Mirage and Nesher Aces* (Oxford: Osprey Publishing, 2004), pp. 82–83.

15 Bar-Siman-Tov, *The Israeli–Egyptian War of Attrition*, pp. 90–102.

16 Shlomo Aloni, *Israeli F-4 Phantom II Aces* (Oxford: Osprey Publishing, 2004), p. 86, and Aloni, *Israeli Mirage and Nesher Aces*, p. 83.

17 In-depth treatments of this bombing campaign may be found in Bar-Siman-Tov, *The Israeli–Egyptian War of Attrition*, pp. 117–173; Yaacov Bar-Siman-Tov, "The Myth of Strategic Bombing: Israeli Deep-Penetration Air Raids in the War of Attrition, 1969–1970," *Journal of Contemporary History*, Vol. 19, No. 3 (July 1984), pp. 549–570; and Avi Shlaim and Raymond Tanter, "Decision, Process, Choice, and Consequences: Israel's Deep-Penetration Bombing in Egypt, 1970," *World Politics*, Vol. 30, No. 4 (July 1978), pp. 483–516.

18 On the Israeli government rationale for deep-penetration bombing, see

Bar-Siman-Tov, *The Israeli–Egyptian War of Attrition*, pp. 120–125, and Shimshoni, *Israel and Conventional Deterrence*, pp. 157–159.

19 On the Israeli government assessment of American and Soviet conduct in respect of deep-penetration bombing, see Bar-Siman-Tov, *The Israeli–Egyptian War of Attrition*, pp. 126–132, and Shimshoni, *Israel and Conventional Deterrence*, p. 159.

20 For the particulars of the deep-penetration bombing, see Bar-Siman-Tov, *The Israeli–Egyptian War of Attrition*, pp. 134–135; the official IAF website, <www.iaf.org.il>; Andreas Klein and Shlomo Aloni, *Israeli Phantoms: The "Kurnass" in IDF/AF Service, 1969–1988* (Erlangen: Double Ugly! Books, 2009), pp. 31–35; and Danny Shalom, *Phantoms Over Cairo: The Israeli Air Force in the War of Attrition (1967–1970)*, Vol. B (Rishon Lezion: Bavir Aviation & Space Publications, 2007).

21 Two other squadrons carried out a total of three raids on January 13, the only date during the entirety of Blossom when attacks were carried out by additional squadrons. Still, other squadrons participated on the fringes of Blossom—such as in the combat air patrol (CAP), communications relay, and search-and-rescue (SAR) roles—but they did not participate in the bombing itself.

22 David A. Korn, *Stalemate: The War of Attrition and Great Power Diplomacy in the Middle East, 1967–1970* (Boulder, CO: Westview Press, 1992), p. 177.

23 On the Khanka and Ma'adi raids, see Bar-Siman-Tov, *The Israeli–Egyptian War of Attrition*, p. 134.

24 Ibid., pp. 134–135.

25 Ibid., p. 134.

26 Shimshoni, *Israel and Conventional Deterrence*, p. 160. During two raids, IAF aircraft also inadvertently inflicted serious collateral damage. In the first instance, one aircraft dropped its bombs on a factory that closely resembled its intended target while dodging antiaircraft fire. In the second instance, one aircraft dropped its bombs on a school located within the perimeter of an army base On these incidents, see Korn, *Stalemate*, pp. 182–183 and 196–197.

27 Isabella Ginor and Gideon Remez, *The Soviet–Israeli War 1967–1973: The USSR's Military Intervention in the Egyptian–Israeli Conflict* (Oxford: Oxford University Press, 2017), p. 144.

28 Klein and Aloni, *Israeli Phantoms*, pp. 32–35.

29 Bar-Siman-Tov, *The Israeli–Egyptian War of Attrition*, pp. 139–144, and Shimshoni, *Israel and Conventional Deterrence*, pp. 161–165.

30 The connection between American arms transfers and Israeli military policy during the War of Attrition is examined in David Rodman, "Arms and Influence: The American–Israeli Relationship and the 1969–1970 War of Attrition," *Israel Affairs*, Vol. 25, No. 1 (February 2019), pp. 26–

41. While the United States agreed to replace IAF losses on a one-for-one basis, it declined to approve the transfer of additional aircraft.

31 Soviet participation in the War of Attrition, both before and after the deep-penetration bombing, is chronicled exhaustively in Ginor and Remez, *The Soviet–Israeli War*, especially pp. 101–209.

32 It must be conceded, however, that the threat of nuclear annihilation had a major impact on the decision of the Japanese regime to end the war in the Pacific.

33 On Egyptian encouragement of Soviet intervention, see Bar-Siman-Tov, *The Israeli–Egyptian War of Attrition*, pp. 136–139, and Ginor and Remez, *The Soviet–Israeli War*, especially pp. 135–160.

34 On the role of Unit 848, see Dmitry Adamsky, "Israel Intelligence and Soviet Military Involvement in the Middle East," pp. 180–181, in Amos Gilboa and Ephraim Lapid (eds.), *Israel's Silent Defender: An Inside Look at Sixty Years of Israeli Intelligence* (Jerusalem: Gefen Publishing House, 2012).

35 Ibid., p. 182. Parenthetically, the intelligence community also failed to assess properly Soviet behavior on the eve of the Yom Kippur War.

36 Eban, *Personnel Witness*, p. 484.

37 Klein and Aloni, *Israeli Phantoms*, p. 31.

38 Ibid.

39 Quoted in Bar-Siman-Tov, *The Israeli–Egyptian War of Attrition*, pp. 140–141.

40 For these numbers, see Bar-Siman-Tov, *The Israeli–Egyptian War of Attrition*, p. 160; Jon D. Glassman, *Arms for the Arabs: The Soviet Union and War in the Middle East* (Baltimore: Johns Hopkins University Press, 1975), p. 75; and Efraim Karsh, *The Cautious Bear: Soviet Military Engagement in Middle Eastern Wars in the Post-1967 Era* (Boulder, CO: Westview Press, 1987), pp. 16, 64–65, and 77.

41 Shlomo Aloni, *Israeli F-4 Phantom II Aces*, p. 86, and Aloni, *Israeli Mirage and Nesher Aces*, pp. 83–84. On the Soviet–Israeli air clash, see Ginor and Remez, *The Soviet–Israeli War*, pp. 190–194.

42 Simcha Salach (ed.), *Israel Air Force Milestones: Heroes and Legends* (Herzliya: Israel Air Force Center Foundation, n.d.), p. 16.

43 Bar-Siman-Tov, *The Israeli–Egyptian War of Attrition*, pp. 159–169, and Shimshoni, *Israel and Conventional Deterrence*, pp. 165–168. The most thorough account of Soviet participation in the fighting is Ginor and Remez, *The Soviet–Israeli War*, especially pp. 161–197.

44 For IAF losses, see Eliot A. Cohen and John Gooch, *Military Misfortunes: The Anatomy of Failure in War* (New York: The Free Press, 1990), p. 99.

45 Eban, *Personal Witness*, p. 489.

46 On these violations, see Ginor and Remez, *The Soviet–Israeli War*, pp. 199–209, and Shimshoni, *Israel and Conventional Deterrence*, p. 170.

## 5 The Battle of October 14

1 The force ratios on the Golan are enumerated in Dani Asher et al., *Syrians at the Border: Strategies, Tactics, Battles, Israel's Northern Command, 1973* (Jerusalem: Contento de Semrik, 2014), pp. 39–53 and 77–109; Central Intelligence Agency, Directorate of Intelligence, *The 1973 Arab–Israeli War: Overview and Analysis of the Conflict* (Langley, VA: Central Intelligence Agency, 1975), p. 53; Chaim Herzog, *The War of Atonement: October, 1973* (Boston: Little, Brown and Co., 1975), pp. 60–63; and Abraham Rabinovich, *The Yom Kippur War: The Epic Encounter that Transformed the Middle East* (New York: Schocken, 2004), p. 142.

2 The force ratios in the Sinai are enumerated in Directorate of Intelligence, *The 1973 Arab–Israeli War*, pp. 74–78; Herzog, *The War of Atonement*, pp. 150–151; Rabinovich, *The Yom Kippur War*, p. 55; and Emanuel Sakal, *Soldier in the Sinai: A General's Account of the Yom Kippur War* (Lexington, KY: University Press of Kentucky, 2014), p. 77.

3 For United States pressure on Israel to eschew a preemptive strike, see the October 6, 1973 telegram from Secretary of State Henry Kissinger to President Richard Nixon in Middle East War Memos and Miscellaneous (October 1–October 17), NSC File, Nixon Presidential Materials, Box 664, National Archives, and the October 7, 1973 memorandum of conversation between Kissinger and Israeli Ambassador to the United States Simcha Dinitz in Records of Henry Kissinger, 1973–1977, Record Group 59, State Department Records, Box 25 (Category C 1973 Arab–Israeli War), National Archives. Also see Michael Brecher, *Decisions in Crisis: Israel, 1967 and 1973* (Berkeley: University of California Press, 1980), pp. 177–178, 187–188, and 197–201.

4 The nature and purpose of the Bar Lev Line are examined in Simon Dunstan, *Israeli Fortifications of the October War 1973* (Oxford: Osprey Publishing, 2008).

5 For an exploration of the reasons behind the IDF's more rapid turnaround on the Golan than in the Sinai, see David Rodman, "A Tale of Two Fronts: Israeli Military Performance during the Early Days of the 1973 Yom Kippur War," *The Journal of Military History*, Vol. 82, No. 1 (January 2018), pp. 208–218.

6 Evaluations of the October 8 counterattack in the Sinai are found in John J. McGrath, "Sinai 1973: Israeli Maneuver Organization and the Battle of the Chinese Farm," pp. 67–68, in John J. McGrath (ed.), *An Army at War: Change in the Midst of Conflict* (Fort Leavenworth, KS: U.S. Army Command and General Staff College, 2005) and, in much more depth, in Sakal, *Soldier in the Sinai*, pp. 208–284.

7 On Israeli losses during the early days of the war, see the October 9, 1973 memorandum of conversation between American and Israeli diplomats in Records of Henry Kissinger, Record Group 59, State Department Records, Lot 91 D 414, Box 25, Arab–Israeli War, National Archives.

These figures include damaged tanks and aircraft that the IDF retrieved, repaired, and returned to service.

8 For Israeli tank losses in the Sinai during the first day of war, see Rabinovich, *The Yom Kippur War*, p. 233, and Sakal, *Soldier in the Sinai*, p. 174.

9 On the insufficient number of tank transporters in the IDF inventory, see Sakal, *Soldier in the Sinai*, p. 102.

10 For Israeli tank losses during the October 8 counterattack, see Jacob Even and Simcha B. Maoz, *At the Decisive Point in the Sinai: Generalship in the Yom Kippur War* (Lexington, KY: University Press of Kentucky, 2017), p. 39, and Sakal, *Soldier in the Sinai*, p. 209.

11 For Israeli aircraft losses during the first three days of the war, see Ra'anan Weiss, *The Israeli AF in the Yom Kippur War: Facts and Figures* (Kfar Tavor: Isradecal Publications, 2014), pp. 33, 47, and 57.

12 Directorate of Intelligence, *The 1973 Arab–Israeli War*, p. 89; Even and Maoz, *At the Decisive Point*, pp. 42–75; and Sakal, *Soldier in the Sinai*, pp. 287–301.

13 Sakal, *Soldier in the Sinai*, p. 300.

14 McGrath, "Sinai 1973," p. 69, and Sakal, *Soldier in the Sinai*, p. 300.

15 For this debate, see Even and Maoz, *At the Decisive Point*, pp. 65–75, and Sakal, *Soldier in the Sinai*, pp. 299–300.

16 On Israel's acquiescence to a cease-fire, see the October 23, 1973 briefing by Kissinger to his staff in Transcripts of Secretary of State Kissinger's Staff Meetings, 1973–77, Record Group 59, Department of State Records, Box 1 (Secretary's Analytical Staff Meetings), National Archives.

17 For Egyptian casualties during the canal crossing, see Herzog, *The War of Atonement*, p. 155, and Kenneth M. Pollack, *Arabs at War: Military Effectiveness, 1948–1991* (Lincoln: University of Nebraska Press, 2002), p. 111.

18 On the Egyptian army's accomplishments during the first days of the war, see Herzog, *The War of Atonement*, pp. 150–192; Pollack, *Arabs at War*, pp. 108–114; and Rabinovich, *The Yom Kippur War*, pp. 101–141 and 235–252.

19 Weiss, *The Israeli AF*, p. 56.

20 For this school of thought, see Dani Asher, *The Egyptian Strategy for the Yom Kippur War: An Analysis* (Jefferson, NC: McFarland & Company, Inc., 2009), pp. 56–62, and Pollack, *Arabs at War*, pp. 114–116 and 603–604, n. 165.

21 For a succinct discussion of the influence of Soviet doctrine on Egyptian war planning before the Yom Kippur War, see Avraham Adan, *On the Banks of the Suez: An Israeli General's Personal Account of the Yom Kippur War* (Novato, CA: Presidio Press 1980), pp. 62–68.

22 Adan, *On the Banks of the Suez*, p. 236.

23 Herzog, *The War of Atonement*, p. 203, and Rabinovich, *The Yom Kippur War*, p. 346.
24 The Egyptian battle plan is compactly enumerated in Pollack, *Arabs at War*, p. 116.
25 The Egyptian ORBAT is pieced together from Even and Maoz, *At the Decisive Point*, p. 77; Pollack, *Arabs at War*, p. 115; and Sakal, *Soldier in the Sinai*, p. 303.
26 The Israeli ORBAT is pieced together from Even and Maoz, *At the Decisive Point*, pp. 76–77; Pollack, *Arabs at War*, p. 115; and Sakal, *Soldier in the Sinai*, p. 303.
27 The Battle of October 14 is chronicled by, inter alia, Adan, *On the Banks of the Suez*, pp. 236–242; Hanoch Bartov, *Dado: 48 Years and 20 Days* (Tel Aviv: Ma'ariv Book Guild, 1981), pp. 460–470; Directorate of Intelligence, *The 1973 Arab–Israeli War*, pp. 89–90; Even and Maoz, *At the Decisive Point*, pp. 76–83; Herzog, *The War of Atonement*, pp. 205–207; Pollack, *Arabs at War*, pp. 114–117; Rabinovich, *The Yom Kippur War*, pp. 352–356; Sakal, *Soldier in the Sinai*, pp. 302–304; and Ariel Sharon (with David Chanoff), *Warrior: An Autobiography* (New York: Simon & Schuster, 1989), pp. 310–311. The following description of the battle and its results is compiled from these sources.
28 Ofer Aderet, "The Israeli Tank Troops Who Destroyed an Egyptian Brigade in the Yom Kippur War," *Ha'aretz*, October 2, 2017 (Internet edition).
29 Ibid.
30 Israeli fighter-bombers, on the other hand, did not actually destroy many Egyptian tanks, as the IAF's capability to knock out armored vehicles during the Yom Kippur War was quite limited. The claim by Even and Maoz, *At the Decisive Point*, p. 79, to the effect that the IAF destroyed a total of 20 tanks taking part in the 3rd (EG) Army Corps central and southern thrusts would appear to be a considerable exaggeration.
31 Anthony H. Cordesman and Abraham R. Wagner, *The Lessons of Modern War: The Arab–Israeli Conflicts, 1973–1989*, Vol. 1 (Boulder, CO: Westview Press, 1990), pp. 49–50.
32 Samuel M. Katz, *Follow Me!: A History of Israel's Military Elite* (London: Arms and Armour Press, 1989), p. 125, and Pollack, *Arabs at War*, p. 116.
33 On this Egyptian source and his contributions to Israeli national security, see Uri Bar-Joseph, *The Angel: The Egyptian Spy Who Saved Israel* (New York: HarperCollins, 2016).
34 Rabinovich, *The Yom Kippur War*, pp. 345–346.
35 For specific examples of the importance of reconnaissance forces on October 14, see Adan, *On the Banks of the Suez*, p. 238, and Rabinovich, *The Yom Kippur War*, p. 355.
36 Sharon, *Warrior*, pp. 307–308.
37 Pollack, *Arabs at War*, p. 604, n. 166, and Sakal, *Soldier in the Sinai*, p. 304.

38 Adan, *On the Banks of the Suez*, p 238, and Even and Maoz, *At the Decisive Point*, p. 82.
39 Adan, *On the Banks of the Suez*, p. 240. Adan, the commander of the 162nd (IS) Armored Division in the Yom Kippur War, attributes the losses in part to this unfamiliarity.
40 Cordesman and Wagner, *The Lessons of Modern War*, p. 61.
41 The IDF did not receive modern American antitank missiles until the last days of the war. Ibid., pp. 64–65.
42 Ibid., pp. 66–68.
43 Ibid., p. 99.
44 Adan, *On the Banks of the Suez*, p. 241, and Sakal, *Soldier in the Sinai*, p. 304.
45 Sakal, *Soldier in the Sinai*, pp. 111–323, and Even and Maoz, *At the Decisive Point*, pp. 13–75, remark extensively (and bitterly) on the $C^2$ dysfunction within IDF forces in the Sinai throughout the first week of hostilities. Sakal served for much of the war as a tank battalion commander in the 401st (IS) Armored Brigade of the 252nd (IS) Armored Division. Even served as the deputy division commander of the 143rd (IS) Armored Division, and Maoz served as Even's assistant.
46 On $C^2$ problems during the Israeli counteroffensive, see Amiram Ezov, *Crossing: Suez, 1973* (Tel Aviv, ContentoNow, 2016). Cf. also the works of Adan, Bartov, Even and Maoz, Herzog, Rabinovich, and Sharon listed in note 27.
47 Oakland McCulloch, *The Decisiveness of Israeli Small-Unit Leadership on the Golan Heights in the 1973 Yom Kippur War*, Master's Thesis (Fort Leavenworth, KS: United States Army Command and General Staff College, 2003), p. 77. An online copy of this thesis can be found at <www.dtic.mil>.
48 Pollack, *Arabs at War*, pp. 116–117.
49 Adan, *On the Banks of the Suez*, p. 237. Also on this point, see Sakal, *Soldier in the Sinai*, p. 304.
50 Pollack, *Arabs at War*, pp. 116–117, and Rabinovich, *The Yom Kippur War*, pp. 354–355.
51 Herzog, *The War of Atonement*, p. 270.
52 On the role of Israeli airpower on October 14, see Adan, *On the Banks of the Suez*, p. 238; Even and Maoz, *At the Decisive Point*, pp. 78–79; Rabinovich, *The Yom Kippur War*, p. 355; and Sakal, *Soldier in the Sinai*, p. 303.
53 Pollack, *Arabs at War*, p. 117.
54 Sakal, *Soldier in the Sinai*, p. 304.
55 Even and Maoz, *At the Decisive Point in the Sinai*, pp. 81–82.
56 Quoted in Herzog, *The War of Atonement*, p. 206. Rabinovich, *The Yom Kippur War*, p. 356, presents a slightly different formulation of this officer's remark, but the substance of the comment is the same.

## 6 Operation Mole Cricket 19

1 Air superiority is achieved when control over a specified airspace is obtained and maintained such that friendly aircraft can execute their missions in that airspace, while enemy aircraft cannot do so.

2 CAS strikes are attacks against enemy forces (and/or assets) in immediate proximity to friendly forces, while interdiction strikes are attacks against enemy forces (and/or assets) not in immediate proximity to friendly forces. Strategic attacks are attacks against high-value targets whose destruction confers more than a local, tactical advantage to friendly forces.

3 An IADS—such as the ones deployed by the Egyptian and Syrian armies in the Yom Kippur War—links surface-to-air missile (SAM) batteries, antiaircraft artillery (AAA) batteries, search and fire-control radars, and a $C^2$ network into a seamless web designed to deny an enemy air force the use of the airspace that the IADS is intended to dominate.

4 On the ultimately futile effort of the IAF to overcome the Soviet–Egyptian IADS in order to retain air superiority over the Suez Canal zone during the War of Attrition, see Yaacov Bar-Siman-Tov, *The Israeli–Egyptian War of Attrition, 1969–1970: A Case Study of Limited Local* War (New York: Columbia University Press, 1980), pp. 145–173, and Isabella Ginor and Gideon Remez, *The Soviet–Israeli War, 1967–1973: The USSR's Military Intervention in the Egyptian–Israeli Conflict* (New York: Oxford University Press, 2017), pp. 135–209.

5 For the specifics behind Operations Challenge 4 and Model 5, see Ra'anan Weiss, *The Israeli AF in the Yom Kippur War: Facts and Figures* (Kfar Tavor: Isradecal, 2014), p. 38.

6 In a personal conversation with the author at the Israel Air Force Center (IAFC), Hezliya, Israel, on June 25, 2007, Brigadier General (Ret.) Itzhak Amitay, a veteran of many missions on the Sinai front during the war, attested to the fact that the IAF was indeed making progress in implementing Challenge 4 before IDF GHQ scrubbed the operation. On this point, also see Abraham Rabinovich, *The Yom Kippur War: The Epic Encounter That Transformed the Middle East* (New York: Schocken, 2004), pp. 175–180, and Emanuel Sakal, *Soldier in the Sinai: A General's Account of the Yom Kippur War* (Lexington, KY: University Press of Kentucky, 2014), pp. 338–342.

7 Anthony H. Cordesman and Abraham R. Wagner, *The Lessons of Modern War: The Arab–Israeli Conflicts, 1973–1989*, Vol. 1 (Boulder, CO: Westview Press, 1989), p. 83, and Weiss, *The Israeli AF*, p. 38.

8 The performance of the IAF in the Yom Kippur War is reviewed in David Rodman, *Israel in the 1973 Yom Kippur War: Diplomacy, Battle, and Lessons* (Brighton: Sussex Academic Press, 2017), pp. 76–80.

9 Quoted in Rebecca Grant, "The Bekaa Valley War," *Air Force Magazine*, Vol. 85, No. 6 (June 2002), p. 59.

10 For a perspective that lays the blame for the IAF's less-than-decisive

SEAD campaign mainly on poor Israeli political and military decision making early in the Yom Kippur War, see Shmuel L. Gordon, "Air Superiority in the Israel–Arab Wars, 1967–1982," pp. 142–146, in John Andreas Olsen (ed.), *A History of Air Warfare* (Washington, D.C.: Potomac Books, 2010).

11 On the IAF's limited capabilities in respect of SEAD during the war, see Cordesman and Wagner, *The Lessons of Modern War*, pp. 82–85.

12 Gordon, "Air Superiority," p. 151.

13 The different components of the air force's SEAD package are cited in Cordesman and Wagner, *The Lessons of Modern War*, pp. 186–193; Grant, "The Bekaa Valley War," pp. 60–62; Gordon, "Air Superiority," p. 149; and Benjamin S. Lambeth, *Moscow's Lessons from the 1982 Lebanon Air War* (Santa Monica, CA: The Rand Corporation, 1984), pp. 5–8.

14 On the IAF training effort, see Cordesman and Wagner, *The Lessons of Modern War*, p. 190; Grant, "The Bekaa Valley War," p. 60; and Lambeth, *Moscow's Lessons*, p. 5.

15 Cordesman and Wagner, *The Lessons of Modern War*, p. 190.

16 On hostilities between the IDF and the PLO across the Israel–Lebanon border after the Yom Kippur War, see Daniel Byman, *A High Price: The Triumphs and Failures of Israeli Counterterrorism* (New York: Oxford University Press, 2011), pp. 58–65, and Gunther E. Rothenberg, "Israeli Defense Forces and Low-intensity Operations," pp. 69–70, in David Charters and Maurice Tugwell (eds.), *Armies in Low-intensity Conflict: A Comparative Analysis* (London: Brassey's Defense Publishers, 1989).

17 For this string of terrorist attacks, see Byman, *A High Price*, p. 60, and Rothenberg, "Israeli Defense Forces," p. 69.

18 The sequence of military and diplomatic developments in Lebanon in the spring–summer of 1981 is enumerated in Eyal Zisser, "The 1982 'Peace for Galilee War': Looking Back in Anger—Between an Option of a War and a War of No Option," p. 199, in Mordechai Bar-On, *A Never-ending Conflict: A Guide to Israeli Military History* (Westport, CT: Praeger, 2004).

19 The lead-up to the war is chronicled in Ze'ev Schiff and Ehud Ya'ari, *Israel's Lebanon War* (New York: Simon & Schuster, 1984).

20 The behind-the-scenes intrigue within the Israeli government that led to the transformation of Operation Peace for Galilee from a limited operation against the PLO to a full-scale war with both the PLO and Syria lies beyond the scope of a description and analysis of Mole Cricket 19. For an authoritative treatment of this intrigue, see Schiff and Ya'ari, *Israel's Lebanon War*.

21 Accounts of Mole Cricket 19 are found in Eliezer Cohen, *Israel's Best Defense: The First Full Story of the Israeli Air Force* (New York: Orion Books, 1993), pp. 464–472; Cordesman and Wagner, *The Lessons of Modern War*, pp. 186–203; Trevor N. Dupuy and Paul Martell, *Flawed Victory: The Arab–Israeli Conflict and the 1982 War in Lebanon* (Fairfax,

VA: HERO Books, 1986), pp. 119–121 and 144–146; Gordon, "Air Superiority," pp. 148–152; Grant, "The Bekaa Valley War," pp. 60–62; Lambeth, *Moscow's Lessons*, pp. 5–11; and Kenneth M. Pollack, *Arabs at War: Military Effectiveness, 1948–1991* (Lincoln, NE: University of Nebraska Press, 2002), pp. 532–534. The present account of the operation is compiled on the basis of the most authoritative information extracted from these sources.

22 Cordesman and Wagner, *The Lessons of Modern War*, p. 193; Lambeth, *Moscow's Lessons*, p. 7; and Pollack, *Arabs at War*, pp. 532–533.

23 Cordesman and Wagner, *The Lessons of Modern War*, p. 153; Lambeth, *Moscow's Lessons*, p. 11; and Pollack, *Arabs at War*, p. 533.

24 For the number of SEAD sorties, see Gordon, "Air Superiority," p. 151.

25 Accounts of this counter-air effort are found in Cordesman and Wagner, *The Lessons of Modern War*, pp. 193–203; Gordon, "Air Superiority," pp. 151–152; and Lambeth, *Moscow's Lessons*, pp. 8–10.

26 Cordesman and Wagner, *The Lessons of Modern War*, p. 202; Gordon, "Air Superiority," p. 151; Lambeth, *Moscow's Lessons*, p. 9; and Pollack, *Arabs at War*, p. 533. Precise figures in regard to Syrian aircraft losses throughout the Lebanon War vary somewhat from source to source, but the general picture of completely one-sided air battles is made clear in all of them.

27 Cordesman and Wagner, *The Lessons of Modern War*, p. 202, and Gordon, "Air Superiority," p. 152.

28 Cohen, *Israel's Best Defense*, p. 482, and Cordesman and Wagner, *The Lessons of Modern War*, p. 202.

29 Cordesman and Wagner, *The Lessons of Modern War*, p. 189.

30 Ibid, pp. 189–190.

31 Ibid, p. 190.

32 Pollack, *Arabs at War*, p. 532.

33 On the Israeli intelligence-gathering effort, see Cordesman and Wagner, *The Lessons of Modern War*, p. 189, and Lambeth, *Moscow's Lessons*, p. 5.

34 Pollack, *Arabs at War*, pp. 542–543.

35 Cordesman and Wagner, *The Lessons of Modern War*, p. 192.

36 Pollack, *Arabs at War*, p. 532.

37 Cohen, *Israel's Best Defense*, p. 468.

38 On the IAF ORBAT, see Cordesman and Wagner, *The Lessons of Modern War*, pp. 186–203 and 206–210.

39 A number of other, considerably older and less advanced, SAM batteries (thought by observers to be) located inside Syria proper supported the IADS in the Bekaa, but these batteries were apparently not directly connected to it. For the composition of the Syrian IADS, see Cordesman and Wagner, *The Lessons of Modern War*, p. 193; Gordon, "Air Superiority," pp. 149–150; Grant, "The Bekaa Valley War," p. 60; and Lambeth, *Moscow's Lessons*, pp. 6–7.

40 Grant, "The Bekaa Valley War," pp. 60–61.
41 Lambeth, *Moscow's Lessons*, p. 8.
42 Cordesman and Wagner, *The Lessons of Modern War*, p. 187.
43 Pollack, *Arabs at War*, p. 532.
44 Ibid., p. 533. Also see Cordesman and Wagner, *The Lessons of Modern War*, p. 200, and Lambeth, *Moscow's Lessons*, p. 8.
45 Pollack, *Arabs at War*, p. 533.
46 On IAF performance in the CAS and interdiction roles during the Lebanon War, see Cordesman and Wagner, *The Lessons of Modern War*, pp. 203–206.
47 Ibid, p. 143.
48 On the performance of Israeli antitank helicopters during the Lebanon War, see Cohen, *Israel's Best Defense*, pp. 478–482, and Cordesman and Wagner, *The Lessons of Modern War*, pp. 211–212.
49 Cordesman and Wagner, *The Lessons of Modern War*, pp. 203–206, and Pollack, *Arabs at War*, p. 547.
50 For an insightful treatment of the Israeli role in the current RMA, see Dima Adamsky, *The Culture of Military Innovation: The Impact of Cultural Factors on the Revolution in Military Affairs in Russia, the US, and Israel* (Palo Alto: Stanford University Press, 2010), pp. 93–129.
51 Quoted in Lothar Ibrügger, *The Revolution in Military Affairs* (Brussels: NATO Parliamentary Assembly, 1998). This document may be found at The Information Warfare Site, <www.iwar.org.uk>.
52 Adamasky, *The Culture*, p. 95.
53 Gordon, "Air Superiority," p. 151.
54 Grant, "The Bekaa Valley War," p. 62.
55 Grant, "The Bekaa Valley War," p. 62, and Vladimir Voronov, *The Syrian Nemesis* (London: The Henry Jackson Society, 2017), p. 3.

## 7 Operations Hell and Spring of Youth

1 For a comprehensive history of Israel's national security doctrine, see Charles D. Freilich, *Israeli National Security: A New Strategy for an Era of Change* (New York: Oxford University Press, 2018). Also see David Rodman, "Israel's Post-1948 Security Experience" in Stuart A. Cohen and Aharon Klieman (eds.), *Routledge Handbook on Israeli Security* (London: Routledge, 2019).
2 Accounts of Arab–Israeli hostilities during these periods can be found in Daniel Byman, *A High Price: The Triumphs and Failures of Israeli Counterterrorism* (New York: Oxford University Press, 2011), pp. 18–26; Ami Gluska, "'The War over the Water' During the 1960s" in Mordechai Bar-On (ed.) *A Never-ending Conflict: A Guide to Israeli Military History* (Westport, CT: Praeger, 2004); Gunther E. Rothenberg, "Israeli Defense Forces and Low-intensity Operations," pp. 55–59, in David Charters and Maurice Tugwell (eds.), *Armies in Low-intensity Conflict: A Comparative*

*Analysis* (London: Brassey's Defense Publishers, 1989); and Jonathan Shimshoni, *Israel and Conventional Deterrence: Border Warfare from 1953 to 1970* (Ithaca, NY: Cornell University Press, 1988), pp. 34–122.

3 On the pre-1967 PLO insurgency/terrorist campaign, see Byman, *A High Price*, pp. 32–33, and Rothenberg, "Israeli Defense Forces," pp. 60–61.

4 On the post-1967 PLO insurgency/terrorist campaign, see Byman, *A High Price*, pp. 34–54, and Rothenberg, "Israeli Defense Forces," pp. 61–67.

5 Descriptions of Operation Hell are found in the following: Trevor N. Dupuy, *Elusive Victory: The Arab–Israeli Wars, 1947–1974* (New York: Harper & Row, 1978), pp. 350–356; Samuel M. Katz, *Follow Me!: A History of Israel's Military Elite* (London: Arms & Armour Press, 1989), pp. 93–95; and Kenneth M. Pollack, *Arabs at War: Military Effectiveness, 1948–1991* (Lincoln, NE: University of Nebraska Press, 2002), pp. 330–335. For a personal account of the fighting from an Israeli officer who participated in the battle, see Muki Betser (with Robert Rosenberg), *Secret Soldier: The True Life Story of Israel's Greatest Commando* (New York: The Atlantic Monthly Press, 1996), pp. 6–22. The following account of the operation is compiled from these sources.

6 For the Israeli objective(s), see Dupuy, *Elusive Victory*, pp. 352 and 355–356, and Pollack, *Arabs at War*, pp. 331–332.

7 On the IDF ORBAT, see Dupuy, *Elusive Victory*, p. 350; Katz, *Follow Me!*, p. 93; and Pollack, *Arabs at War*, p. 331.

8 On the PLO and Arab Legion ORBATs, see Dupuy, *Elusive Victory*, pp. 350–351, and Pollack, *Arabs at War*, p. 332.

9 For the Arab portrayal of the results of the battle, see Dupuy, *Elusive Victory*, pp. 354–356; Katz, *Follow Me!*, p. 95; and Pollack, *Arabs at War*, pp. 333–334.

10 IDF, Arab Legion, and PLO losses are noted in Dupuy, *Elusive Victory*, p. 354; Katz, *Follow Me!*, p. 94; and Pollack, *Arabs at War*, p. 334.

11 Katz, *Follow Me!*, pp. 94–95.

12 Pollack, *Arabs at War*, p. 334.

13 Dupuy, *Elusive Victory*, p. 350.

14 On Arab Legion intelligence efforts, see Dupuy, *Elusive Victory*, p. 350, and Pollack, *Arabs at War*, pp. 331 and 334.

15 Katz, *Follow Me!*, p. 94.

16 Ibid.

17 Dupuy, *Elusive Victory*, p. 351, and Pollack, *Arabs at War*, pp. 332–334.

18 Katz, *Follow Me!*, p. 94.

19 Ibid.

20 Dupuy, *Elusive Victory*, p. 353, and Pollack, *Arabs at War*, p. 332.

21 Dupuy, *Elusive Victory*, p. 351.

22 Pollack, *Arabs at War*, pp. 334–335.

23 Katz, *Follow Me!*, p. 94.

24 Dupuy, *Elusive Victory*, p. 356, and Katz, *Follow Me!*, p. 95.

25 On the Munich attack, see Byman, *A High Price*, pp. 46–47, and Aaron J. Klein, *Striking Back: The 1972 Munich Olympics Massacre and Israel's Deadly Response* (New York: Random House, 2005).

26 Descriptions of Operation Spring of Youth are found in the following: Ronen Bergman, *Rise and Kill First: The Secret History of Israel's Targeted Assassinations* (New York: Random House, 2018), pp. 166–173; Katz, *Follow Me!*, pp. 112–116; and Klein, *Striking Back*, pp. 157–170. For a personal account of the fighting from an Israeli officer who participated in the battle, see Betser, *Secret Soldier*, pp. 126–144. The following account of the operation is compiled from these sources.

27 Katz, *Follow Me!*, p. 114.

28 Betser, *Secret Soldier*, p. 129.

29 For the pre-operation preparations of one of the SOF teams, see ibid., pp. 128–136.

30 Ibid., pp. 128–129.

31 Katz, *Follow Me!*, p. 115.

32 Ibid.

33 On the offshore CP, see Betser, *Secret Soldier*, p. 138, and Katz, *Follow Me!*, p. 114. On command responsibility in the hands of junior officers, see Betser, *Secret Soldier*, pp. 137–138, and Katz, *Follow Me!*, p. 115.

34 For these attacks, see Byman, *A High Price*, p. 60, and Rothenberg, "Israeli Defense Forces," p. 69.

# Bibliography


## Articles

Aderet, Ofer, "The Israeli Tank Troops Who Destroyed an Egyptian Brigade in the Yom Kippur War," *Ha'artez*, October 2, 2017 (Internet edition).

Bar-Siman-Tov, Yaacov, "The Myth of Strategic Bombing: Israeli Deep-Penetration Air Raids in the War of Attrition, 1969–1970," *Journal of Contemporary History*, Vol. 19, No. 3 (July 1984).

Grant, Rebecca, "The Bekaa Valley War," *Air Force Magazine*, Vol. 85, No. 6 (June 2002).

Oren, Michael B., "Escalation to Suez: The Egypt–Israel Border War, 1949–56," *Journal of Contemporary History*, Vol. 24, No. 2 (April 1989).

Rodman, David, "A Tale of Two Fronts: Israeli Military Performance during the Early Days of the 1973 Yom Kippur War," *The Journal of Military History*, Vol. 82, No. 1 (January 2018).

Rodman, David, "Arms and Influence: The American–Israeli Relationship and the 1969–1970 War of Attrition," *Israel Affairs*, Vol. 25, No. 1 (February 2019).

Shlaim, Avi and Raymond Tanter, "Decision, Process, Choice, and Consequences: Israel's Deep-Penetration Bombing in Egypt, 1970," *World Politics*, Vol. 30, No. 4 (July 1978).

## Books, Reports, and Studies

Adamsky, Dima, *The Culture of Military Innovation: The Impact of Cultural Factors on the Revolution in Military Affairs in Russia, the US, and Israel* (Palo Alto: Stanford University Press, 2010).

Adan, Avraham, *On the Banks of the Suez: An Israeli General's Personal Account of the Yom Kippur War* (Novato, CA: Presidio Press 1980).

Aloni, Shlomo, *Israeli F-4 Phantom II Aces* (Oxford: Osprey Publishing, 2004).

Aloni, Shlomo, *Israeli Mirage and Nesher Aces* (Oxford: Osprey Publishing, 2004).

Aloni, Shlomo, *The June 1967 Six-Day War: Operation Focus*, Vol. A (Bat Hefer: Isradecal, 2008).

Amir, Amos, *Fire in the Sky: Flying in Defence of Israel* (South Yorkshire: Pen and Sword Books, 2005).

Asher, Dani, *The Egyptian Strategy for the Yom Kippur War: An Analysis* (Jefferson, NC: McFarland & Company, Inc., 2009).

Asher, Dani et al., *Syrians at the Border: Strategies, Tactics, Battles, Israel's Northern Command, 1973* (Jerusalem: Contento de Semrik, 2014).

Bar-Joseph, Uri, *The Angel: The Egyptian Spy Who Saved Israel* (New York: HarperCollins, 2016).

Bar-On, Mordechai (ed.), *A Never-ending Conflict: A Guide to Israeli Military History* (Westport, CT: Praeger, 2004).

Bar-Siman-Tov, Yaacov, *The Israeli–Egyptian War of Attrition, 1969–1970: A Case Study of Limited Local War* (New York: Columbia University Press, 1980).

Bartov, Hanoch, *Dado: 48 Years and 20 Days* (Tel Aviv: Ma'ariv Book Guild, 1981).

Beckman, Morris, *The Jewish Brigade: An Army with Two Masters, 1944–45* (New York: De Capo Press, 1998).

Bergman, Ronen, *Rise and Kill First: The Secret History of Israel's Targeted Assassinations* (New York: Random House, 2018).

Betser, Muki (with Robert Rosenberg), *Secret Soldier: The True Life Story of Israel's Greatest Commando* (New York: The Atlantic Monthly Press, 1996).

Brecher, Michael, *Decisions in Crisis: Israel, 1967 and 1973* (Berkeley: University of California Press, 1980).

Byman, Daniel, *A High Price: The Triumphs and Failures of Israeli Counterterrorism* (New York: Oxford University Press, 2011).

Charters, David and Maurice Tugwell (eds.), *Armies in Low-intensity Conflict: A Comparative Analysis* (London: Brassey's Defence Publishers, 1989).

Cohen, Eliezer, *Israel's Best Defense: The First Full Story of the Israel Air Force* (New York: Orion Books, 1993).

Cohen, Eliot A. and John Gooch, *Military Misfortunes: The Anatomy of Failure in War* (New York: The Free Press, 1990).

Cohen, Stuart A. and Aharon Klieman (eds.), *Routledge Handbook on Israeli Security* (London: Routledge, 2019).

Cordesman, Anthony H. and Abraham R. Wagner, *The Lessons of Modern War: The Arab–Israeli Conflicts, 1973–1989*, Vol. 1 (Boulder, CO: Westview Press, 1990).

Dinur, Ben-Zion (ed.), *History of the Hagana*, Vols. 1–3 (Tel Aviv: Am Oved Publishers, 1973).

Dunstan, Simon, *Israeli Fortifications of the October War 1973* (Oxford: Osprey Publishing, 2008).

Dunstan, Simon, *The Six Day War 1967*, Vol. 1 (Oxford: Osprey Publishing, 2009).

Dupuy, Trevor N., *Elusive Victory: The Arab–Israeli Wars, 1947–1974* (New York: Harper & Row, 1978).

Dupuy, Trevor N. and Paul Martell, *Flawed Victory: The Arab–Israeli Conflict and the 1982 War in Lebanon* (Fairfax, VA: HERO Books, 1986).

Eban, Abba, *Personal Witness: Israel Through My Eyes* (New York: G.P. Putnam's Sons, 1984).

Ellis, John, *Brute Force: Allied Strategy and Tactics in the Second World War* (New York: Viking, 1990).

Even, Jacob and Simcha B. Maoz, *At the Decisive Point in the Sinai: Generalship in the Yom Kippur War* (Lexington, KY: University Press of Kentucky, 2017).

Freilich, Charles D., *Israeli National Security: A New Strategy for an Era of Change* (New York: Oxford University Press, 2018).

Gawrych, George W., *Key to the Sinai: The Battles for Abu Ageila in the 1956 and 1967 Arab–Israeli Wars* (Fort Leavenworth, KS: U.S. Army Command and General Staff College, 1990).

Gilboa, Amos and Ephraim Lapid (eds.), *Israel's Silent Defender: An Inside Look at Sixty Years of Israeli Intelligence* (Jerusalem: Gefen Publishing House, 2012).

Ginor, Isabella and Gideon Remez, *The Soviet–Israeli War 1967–1973: The USSR's Military Intervention in the Egyptian–Israeli Conflict* (Oxford: Oxford University Press, 2017).

Glassman, Jon D., *Arms for the Arabs: The Soviet Union and War in the Middle East* (Baltimore: Johns Hopkins University Press, 1975).

Golani, Motti, *Israel in Search of a War: The Sinai Campaign, 1955–1956* (Brighton: Sussex Academic Press, 1998).

Hammel, Eric, *Six Days in June: How Israel Won the 1967 Arab–Israeli War* (New York: Scribner's, 1992).

Hanson, Victor Davis, *Carnage and Culture: Landmark Battles in the Rise of Western Power* (New York: Doubleday, 2001).

Harris, J. P. and F. H. Toase (eds.), *Armoured Warfare* (New York: St. Martin's Press, 1990).

Henkin, Yagil, *The 1956 Suez War and the New World Order in the Middle East: Exodus in Reverse* (Lanham, MD: Lexington Books, 2015).

Herzog, Chaim, *The Arab–Israeli Wars: War and Peace in the Middle East from the War of Independence Through Lebanon*, Rev. ed. (New York: Vintage Books, 1984).

Herzog, Chaim, *The War of Atonement: October, 1973* (Boston: Little, Brown and Co., 1975).

Hoffman, Bruce, *Anonymous Soldiers: The Struggle for Israel, 1917–1947* (New York: Knopf, 2015).

House, Jonathan M., *Combined Arms Warfare in the Twentieth Century* (Lawrence: University Press of Kansas, 2001).

Hughes, Matthew, *Britain's Pacification of Palestine: The British Army, the Colonial State, and the Arab Revolt, 1936–1939* (Cambridge: Cambridge University Press, 2019).

Ibrügger, Lothar, *The Revolution in Military Affairs* (Brussels: NATO Parliamentary Assembly, 1998).

Karsh, Efraim, *The Cautious Bear: Soviet Military Engagement in Middle Eastern Wars in the Post-1967 Era* (Boulder, CO: Westview Press, 1987).

Katz, Samuel M., *Follow Me!: A History of Israel's Military Elite* (London: Arms and Armour Press, 1989).

Klieman, Aaron S., *Israel and the World After 40 Years* (Washington, D.C.: Pergamon-Brassey's International Defense Publishers, 1990).

Klein, Aaron J., *Striking Back: The 1972 Munich Olympics Massacre and Israel's Deadly Response* (New York: Random House, 2005).

Klein, Andreas and Shlomo Aloni, *Israeli Phantoms: The "Kurnass" in IDF/AF Service, 1969–1988* (Erlangen: Double Ugly! Books, 2009).

Kober, Avi, *Practical Soldiers: Israel's Military Thought and Its Formative Factors* (Leiden: Brill, 2015).

Korn, David A., *Stalemate: The War of Attrition and Great Power Diplomacy in the Middle East, 1967–1970* (Boulder, CO: Westview Press, 1992).

Lambeth, Benjamin S., *Moscow's Lessons from the 1982 Lebanon Air War* (Santa Monica, CA: The Rand Corporation, 1984).

Laron, Guy, *The Six-Day War: The Breaking of the Middle East* (New Haven: Yale University Press, 2017).

McCulloch, Oakland, *The Decisiveness of Israeli Small-Unit Leadership on the Golan Heights in the 1973 Yom Kippur War*, Master's Thesis (Fort Leavenworth, KS: United States Army Command and General Staff College, 2003).

McGrath, John J. (ed.), *An Army at War: Change in the Midst of Conflict* (Fort Leavenworth, KS: U.S. Army Command and General Staff College, 2005).

Morris, Benny, *1948: A History of the First Arab–Israeli War* (New Haven: Yale University Press, 2008).

Naor, Moshe, *Social Mobilization in the Arab–Israeli War of 1948: On the Israeli Home Front* (London: Routledge, 2013).

Olsen, John Andreas (ed.), *A History of Air Warfare* (Washington, D.C.: Potomac Books, 2010).

Oren, Michael B., *Six Days of War: June 1967 and the Making of the Modern Middle East* (New York: Oxford University Press, 2002).

Pollack, Kenneth M., *Arabs at War: Military Effectiveness, 1948–2001* (Lincoln, NE: University of Nebraska Press, 2002).

Pollack, Kenneth M., *Armies of Sand: The Past, Present, and Future of Arab Military Effectiveness* (New York: Oxford University Press, 2019).

Rabinovich, Abraham, *The Yom Kippur War: The Epic Encounter that Transformed the Middle East* (New York: Schocken, 2004).

Rodman, David, *Combined Arms Warfare in Israeli Military History: From the War of Independence to Operation Protective Edge* (Brighton: Sussex Academic Press, 2019).

Rodman, David, *Israel in the 1973 Yom Kippur War: Diplomacy, Battle, and Lessons* (Brighton: Sussex Academic Press, 2017).

Rubin, Avshalom, *The Limits of the Land: How the Struggle for the West Bank Shaped the Arab–Israeli Conflict* (Bloomington, IN: Indiana University Press, 2017).

Sakal, Emanuel, *Soldier in the Sinai: A General's Account of the Yom Kippur War* (Lexington, KY: University Press of Kentucky, 2014).

Salach, Simcha (ed.), *Israel Air Force Milestones: Heroes and Legends* (Herzliya: Israel Air Force Center Foundation, n.d.).

Schiff, Ze'ev, *A History of the Israeli Army: 1874 to the Present*, Rev. ed. (New York: Macmillan Publishing Company, 1985).

Schiff, Ze'ev and Ehud Ya'ari, *Israel's Lebanon War* (New York: Simon & Schuster, 1984).

Searle, Alaric, *Armoured Warfare: A Military, Political and Global History* (London: Bloomsbury, 2017).

Shalom, Danny, *Phantoms Over Cairo: The Israeli Air Force in the War of Attrition (1967–1970)*, Vol. B (Rishon Lezion: Bavir Aviation & Space Publications, 2007).

Shamir, Eitan, *Transforming Command: The Pursuit of Mission Command in the U.S., British, and Israeli Armies* (Palo Alto: Stanford University Press, 2011).

Sharon, Ariel (with David Chanoff), *Warrior: An Autobiography* (New York: Simon & Schuster, 1989).

Shimshoni, Jonathan, *Israel and Conventional Deterrence: Border Warfare from 1953 to 1970* (Ithaca, NY: Cornell University Press, 1988).

Voronov, Vladimir, *The Syrian Nemesis* (London: The Henry Jackson Society, 2017).

Weiss, Ra'anan, *The Israeli AF in the Yom Kippur War: Facts and Figures* (Kfar Tavor: Isradecal Publications, 2014).

Winters, Harold A. et al., *Battling the Elements: Weather and Terrain in the Conduct of War* (Baltimore: Johns Hopkins University Press, 1998).

Zadka, Saul, *Blood in Zion: How the Jewish Guerrillas Drove the British Out of Palestine* (London: Brassey's, 1995).

## Primary Sources

Central Intelligence Agency, Directorate of Intelligence, *The 1973 Arab–Israeli War: Overview and Analysis of the Conflict* (Langley, VA: Central Intelligence Agency, 1975).

Nina Howland and Craig Daigle (eds.), *Foreign Relations of the United States, 1969–1976*, Volume XXV, *Arab–Israeli Crisis and War, 1973* (Washington, D.C.: United States Government Printing Office, 2011).

Personal communications with various retired and active IAF officers.

# Index

# About the Author

David Rodman is the author of a number of books about Israeli diplomatic and military history. He has also published articles in peer-reviewed journals, including *War in History*, *Intelligence and National Security*, *The Journal of Military History*, *The Journal of Strategic Studies*, *Middle Eastern Studies*, *Israel Affairs*, *Defence Studies*, and *The Israel Journal of Foreign Affairs*.

Printed and bound by CPI Group (UK) Ltd, Croydon, CR0 4YY

13/09/2026

14961102-0004